AF899963

Bernd-Olaf Küppers

Molecular Theory of Evolution

Outline of a Physico-Chemical Theory of the Origin of Life

Translated by Paul Woolley

With 76 Figures

Springer-Verlag
Berlin Heidelberg New York Tokyo

Bernd-Olaf Küppers
Max-Planck-Institut für Biophysikalische Chemie
Postfach 968, D-3400 Göttingen
Federal Republic of Germany

Translated from the German by Paul Woolley
Max-Planck-Institut für Molekulare Genetik, D-1000 Berlin

Corrected 2nd printing 1985

1st edition 1983

Library of Congress Cataloging in Publication Data
Küppers, Bernd-Olaf, 1944-
Molecular theory of evolution.
Bibliography: p.
Includes index.
1. Chemical evolution. 2. Life--Origin. I. Title.
QH325.K8613 1985 577 85-9800

ISBN-13: 978-3-540-15528-7 e-ISBN-13: 978-3-642-70544-1
DOI: 10.1007/978-3-642-70544-1

This work is subject to copyright. All rights are reserved, whether the whole or part of the material is concerned, specifically those of translation, reprinting, re-use of illustrations, broadcasting, reproduction by photocopying machine or similar means, and storage in data banks. Under § 54 of the German Copyright Law where copies are made for other than private use, a fee is payable to "Verwertungsgesellschaft Wort", Munich.

© Springer-Verlag Berlin Heidelberg 1983

2152/3140-543210

Preface to the Second Impression

Since the first appearance of this book, advances in the analysis of problems of molecular evolution have not been so fundamental as to demand a revision of the text. Therefore, the first edition has been reprinted without any changes except for minor corrections.

Göttingen, June 1985 Bernd-Olaf Küppers

Preface

The subject of this book is the physico-chemical theory of the origin of life. Although this theory is still *in statu nascendi*, it has been developed in recent years to the point where a coherent presentation is possible.

The book is intended as an introductory text for students of physics, chemistry or biology. This interdisciplinary aim has necessitated a choice of material based on the lowest common denominator of physicists and biologists. In particular, the predominantly mathematical and biological sections have been presented at the most elementary level possible. The less mathematically inclined reader can omit Sections (5.3)-(5.5), (5.7)-(5.10), (6.2), (6.3), (9.1)-(9.3), (12.1) and (13.3) without losing the overall view.

For critical reading of the manuscript, for discussions and for many useful suggestions I wish to thank M. Eigen (Göttingen), W.C. Gardiner (Austin), D. Pörschke (Göttingen), P. Schuster (Vienna), P.R. Wills (Auckland) and P. Woolley (Berlin).

The translation of the original, German manuscript into English was kindly undertaken by Paul Woolley. During this and subsequent stages of revision he introduced a great many improvements in the text and the presentation of material. My particular thanks are due to him for his decisive contribution to this book.

Last of all I wish to thank Ingeborg Lechten for typing the text in its various stages of evolution. The completion of this book is largely to be attributed to her patience and efficiency.

Göttingen, September 1982 Bernd-Olaf Küppers

Contents

Introduction

In the course of its history, physics has frequently stepped outside the area of direct experience. It has usually done so in pursuit of one of three goals:

(1) the microcosm,

(2) the macrocosm,

(3) the realm of complexity.

The first two fields have long been part of the established territory of physicists. The third is today still largely virgin ground. Where does the physicist encounter the phenomenon "complexity"?

It frequently comes about that a physical problem is intractable not because of the inadequacy of accepted theory, but rather on account of the complexity of the phenomena under investigation. Much intensive research in physics is moving away from simple systems and towards more complex ones with many variables.

The prime example of a complex system is a living organism. Thus the present monograph has emerged from the activity of a cell population in which around 10^{13} cellular units co-operate in an organized and functionally co-ordinated manner. But even a single cell is built up in a monumentally complex way. In the simplest - *e.g.*, bacterial - cells, more than 10^7 biological macromolecules interact coherently so as to sustain the state of material order which we call "life". Even in the bacterial cell the molecular information-store carries a programme which is only one of more than $10^{2\,000\,000}$ combinatorial possibilities.

These numbers convey a first impression of the kind of complexity with which physics has to deal in considering living systems. This leads to the question of whether the ideas and methods of contemporary physics can conceivably suffice to explain systems of this immense complexity. The great variety of biological systems, along with a proper modesty, forbids a global answer; if we wish to work towards a solution to this problem, we shall have to restrict our consideration to particular biological phenomena.

We intend to investigate here the problem of how life arose. If there should be a fundamental physical difference between living and non-living matter, then the transition from one to the other is the point where the difference must become clearest. However, our main discussion will not be a philosophical pursuit of the reducibility of biology to physics and chemistry, but will bear on the issue itself: the physics of the origin of life.

It is quite clear that a phenomenon like the emergence of living matter possesses many different aspects, so that explicit questions about it will differ greatly in content and emphasis. If we choose to regard the origin of life as a continuous process, we can distinguish at least three phases:

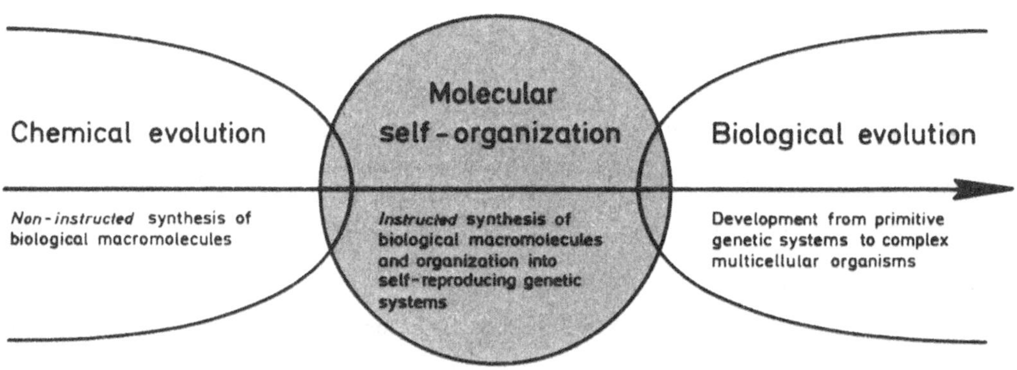

(1) Life could only have arisen in a material medium which had an appropriate chemical composition. There must therefore have been a phase of *chemical evolution*, during which all the chemicals were formed which were needed for the nucleation of life. To these belong the two most important classes of biological macromolecules, the nucleic acids and the proteins. During this initial phase of evolution, the synthesis of biological molecules was *non-instructed*.

(2) As soon as nucleic acids and proteins were present, physical forces between them must have led to a large number of "feedback loops" of mutual interaction, these leading to the *instructed* synthesis of biological macromolecules. The products of this phase of *molecular self-organization* must at some point have begun to resemble the complex, organized units which we see today in the optimized, self-reproducing biosynthetic cycle of the living cell.

(3) In the final phase of the origin of life the divergent process of *biological evolution* commenced, which has resulted in the further development of primitive, single-celled organisms and their differentiation into highly-developed multicellular beings. During this phase of evolution the great diversity of microscopic and macroscopic life appeared.

The actual transition from non-living matter to living is clearly to be attributed to the phase of molecular self-organization, so this is the phase with which we shall be concerned in this monograph. To discuss it we shall proceed according to the following programme:

In Part I we shall summarise the molecular basis of biological information. This is called for by the fact that we can describe the origin of life in a physical way only if we have first formulated the problem in physical terms. At the end of Part I a plausibility argument is used to show that "the origin of life" is equivalent to "the origin of biological information". We shall further see that the origin of biological information is not necessarily to be understood as a unique chance event. On the contrary, if macromolecules are subject to natural selection in the Darwinian sense then a kind of material learning-process takes place, from which the information needed for the construction of a living system necessarily arises.

In Part II the selection principle will be formulated mathematically and its consequences for the evolution of biological macromolecules will be demonstrated. The Darwinian selection principle will be shown to be physically deducible and exactly describable in terms of defined conditions and constraints.

Selection in the classical Darwinian sense is a necessary but not a sufficient condition for the origin of biological information. This will be the main conclusion of Part III. Along with the Darwinian principle there must be a further principle of molecular self-organization which leads to the accumulation of information. The logically simplest realisation of this is the *hypercycle*, proposed by M. Eigen (1971) and developed in detail by M. Eigen and P. Schuster (1979). The properties of the hypercycle occupy a central place in Part III.

However, models and theories are only meaningful and relevant when they proceed from realistic assumptions and well-corroborated experimental data. We shall therefore investigate in Part IV the presuppositions behind the molecular theory of evolution and ways in which this theory may be subjected to experimental test.

I. The Molecular Basis of Biological Information

1. Definition of Living Systems

One of the best-substantiated principles of modern biology is the development of living beings by divergent evolution. A "gradient" of evolution can also be clearly determined: less complex organisms have always given rise to more complex ones. This principle of descendence provides the basis of the biologists' tree of evolution (Fig. 1.1).

The smallest integrated unit of living matter found today is the cell. The fact that the cells of all living beings have the same material basis and are organized according to the same principles is in turn a consequence of the principle of descendence. The question "What is the origin of life?" is thus in this sense equivalent to the question "How did the first cell originate?" To find a physical answer to this, we start by asking what properties are particularly characteristic of a *living* system.

In order to formulate a meaningful definition of a living being we must first know whether we are looking for a sharply-drawn dividing line between non-living and living matter, or whether there is a gradual transition from one to the other. If there is a sharp boundary we can in principle provide a series of necessary *and* sufficient conditions, and thus define a living being unambiguously. If not, we can only state necessary conditions for a system to be alive. Expressed differently: if there is a sharp boundary it should be possible to find properties typical exclusively of living systems. If not, then there is no general criterion, apart from subjective experience, to distinguish the living from the non-living. We could then at best say with certainty of a material system that it is *not* alive, and we should have to state as many necessary conditions as possible, so as to make our definition as narrow as possible.

In order to avoid getting lost in such details before even starting, let us first of all clarify the question "What is life?" with a plausibility argument, and then try to develop a working definition of a living being.

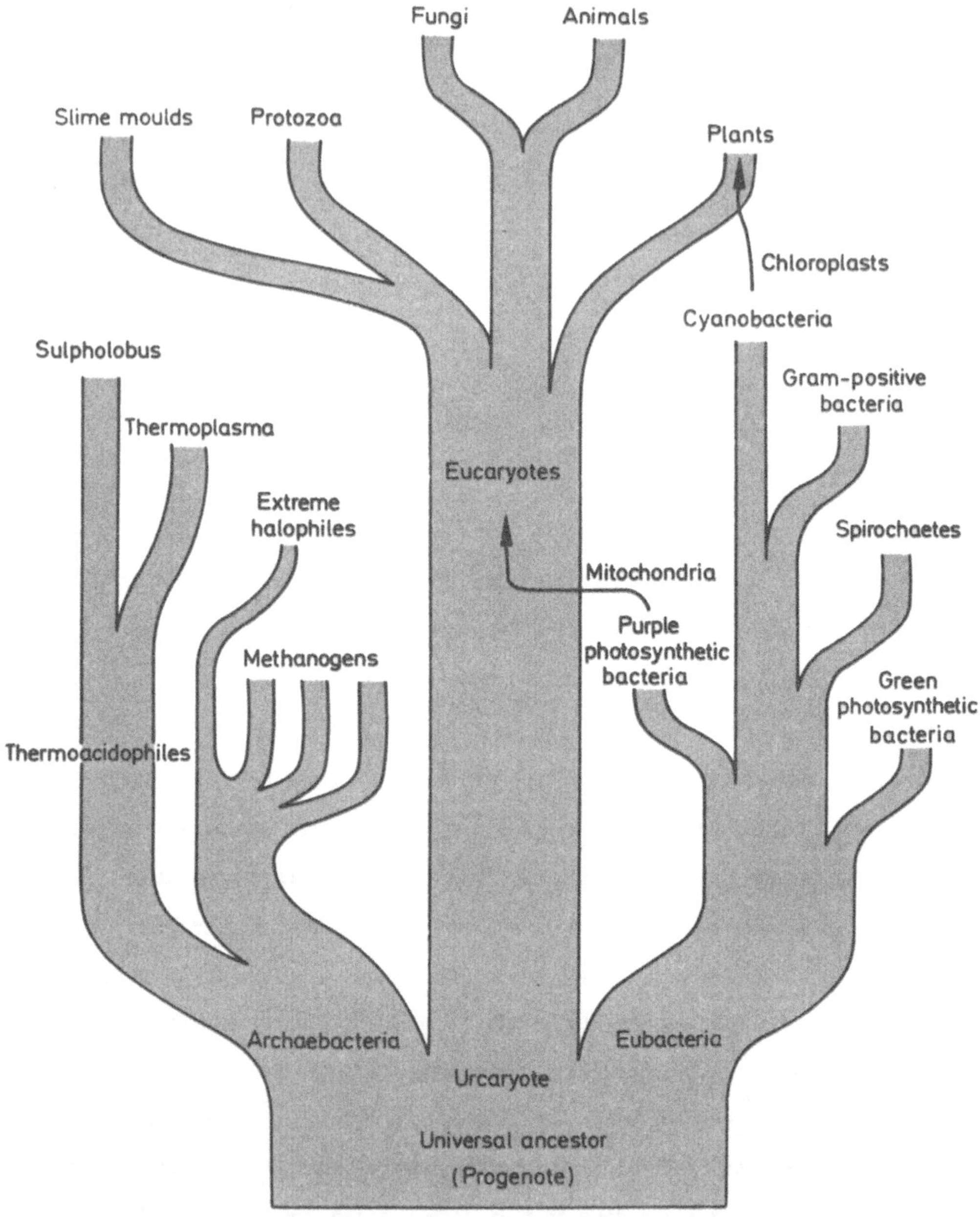

Fig. 1.1. Tree of evolution beginning at the threshold of life. The archaebacteria, the eubacteria and the urcaryote (protocaryote) - the precursor of the eucaryotes - most probably originated from a common ancestor (the progenote) much simpler than the simplest present-day cells. (According to Woese, 1981.)

Modern biology suggests strongly that there is indeed no sharp dividing line between living and non-living systems, but instead a continuous transition. This is supported by the existence of biological objects (*e.g.* viruses), which apparently belong to the border region between inert and living matter (see Chapter 13).

A virus particle has all the properties of a living system when it is inside its host cell, such as a bacterial or a human cell. It exploits the metabolism of its host cell in order to reproduce itself; it mutates; it is subject, within its restricted field of existence, to natural selection. Like every other organism, it has in the course of evolution become adapted to its function, and like every other organism it can only survive under certain environmental conditions.

Thus, within the milieu of its host cell, a virus possesses all the properties that one ordinarily expects of a living being. However, when we remove the virus from this milieu we can crystallize it like any ordinary molecule. In the crystal it behaves in every way like a simple molecule (Klug and Butler, 1978) and fulfils none of the conceivable criteria of being alive.

Without going into the details of biological processes, we can observe in all living organisms the following three properties:

(1) Every living system is characterised by *metabolism*. The presence of metabolism is therefore a necessary condition for a system to be alive. As we shall see, the fundamental reason for this is the complexity of the material structure of every organism. The fact that organisms metabolize means also that they are, in the thermodynamic sense, *open* systems, that is, they exchange matter and energy with their environment.

(2) A second property which we observe in all living beings is the faculty of *self-reproduction*. This too is ultimately a consequence of biological complexity. Organisms are constructed so intricately that they could never arise *de novo* in each generation. If this were possible then there would be no need for self-reproduction.

(3) The third and last property which we see in all living organisms is *mutability*. If organisms only ever made exact copies of themselves there would be no evolutionary progress. It is only the continual variation of organisms, through mutation, which provides Nature with the assortment from which she selects those types best fitted to survive and further reproduce in their milieu.

These considerations lead us to the central postulate that the following properties are necessary for a system to be called "alive":

(1) metabolism,
(2) self-reproduction,
(3) mutability.

These criteria were first set up in 1924 by the Russian biologist A. Oparin, in order to demarcate living from non-living systems. They have sometimes been augmented by a fourth, that of *natural selection*. We shall not include this in our definition of a living being, since it is in fact not an independent criterion. Firstly, the phenomenon of natural selection is a property of a system composed of a *population* of organisms, and secondly, as we shall see, in particular circumstances where metabolism, self-reproduction and mutability all appear, selection in the Darwinian sense takes place automatically.

The three criteria of Oparin are necessary but not sufficient conditions for a system to be called "alive". However they provide a suitable working definition of a primitive organism. They are all connected, as already mentioned, with the complexity of the material structure of living systems. We must therefore now take a closer look at the molecular roots of biological complexity. This analysis will enable us to justify our working definition of living systems in more detail.

2. Structure and Function of Biological Macromolecules

The simplest organisms capable of providing their own metabolism are the bacteria. As single-celled organisms, the bacteria are so to speak "atoms of life", and are therefore excellent model systems for the investigation of all the details important in the molecular structure and function of living organisms.

However, even a bacterial cell is chemically very complex. As Table 1.1 shows, the cell is built up out of numerous small molecules (H_2O, inorganic ions *etc.*) and large ones (macromolecules: proteins, nucleic acids). Under optimal conditions a bacterium such as *Escherichia coli* duplicates itself every twenty minutes. This means that in this relatively short time the entire ordered state of a bacterial cell is reproduced once. The reproduction and the maintenance of biological order, however, require information.

As we shall see in detail, this information can be found in two classes of biological macromolecules: the nucleic acids and the proteins. The macromolecules make a very special contribution to the cellular complexity, for on the one hand they make up a substantial fraction of the cell's mass, and on the other they interact in a highly intricate way and are responsible for the instruction, maintenance and regulation of the various life processes within the system as a whole.

Component	Per cent of total cell weight	Average molecular weight	Approximate number per cell	Number of different kinds
H_2O	70	18	$4 \cdot 10^{10}$	1
Inorganic ions (Na^+, K^+, Mg^{2+}, Ca^{2+}, Fe^{2+}, Cl^-, PO_4^{4-}, SO_4^{2-} *etc.*)	1	40	$2.5 \cdot 10^8$	20
Carbohydrates and precursors	3	150	$2 \cdot 10^8$	200
Amino acids and precursors	0.4	120	$3 \cdot 10^7$	100
Nucleotides and precursors	0.4	300	$1.2 \cdot 10^7$	200
Lipids and precursors	2	750	$2.5 \cdot 10^7$	50
Other small molecules (haem, quinones, breakdown products of food molecules *etc.*)	0.2	150	$1.5 \cdot 10^7$	250
Proteins	15	$4 \cdot 10^4$	10^6	2000 to 3000
Nucleic acids				
DNA	1	$2.5 \cdot 10^9$	4	1
RNA	6			
16S rRNA		$5 \cdot 10^5$	10^5	1
23S rRNA		10^6	10^5	1
tRNA		$2.5 \cdot 10^4$	10^6	60
mRNA		10^5	10^3	10^3

Table 1.1. Approximate chemical composition of a rapidly-growing *Escherichia coli* cell. (After Watson, 1976.)

2.1. The Nucleic Acids

Let us first take a look at the nucleic acids. Their primary function is the storage of biological information. Every organism contains the complete "blueprint" for its own construction, in a package of nucleic acid, a copy of which is present in almost every cell in the organism. In this molecular form the blueprint is handed down from cell to cell and from generation to generation. In order for a molecule to be able to carry all the information necessary to programme the assembly of an organism, it must obviously have quite unusual structural properties.

Today we know in some detail how the molecular script of biology uses an alphabet of only four letters, represented by the four different subunits of a nucleic acid shown schematically in Fig. 2.1. These "building blocks" - the nucleotides - are

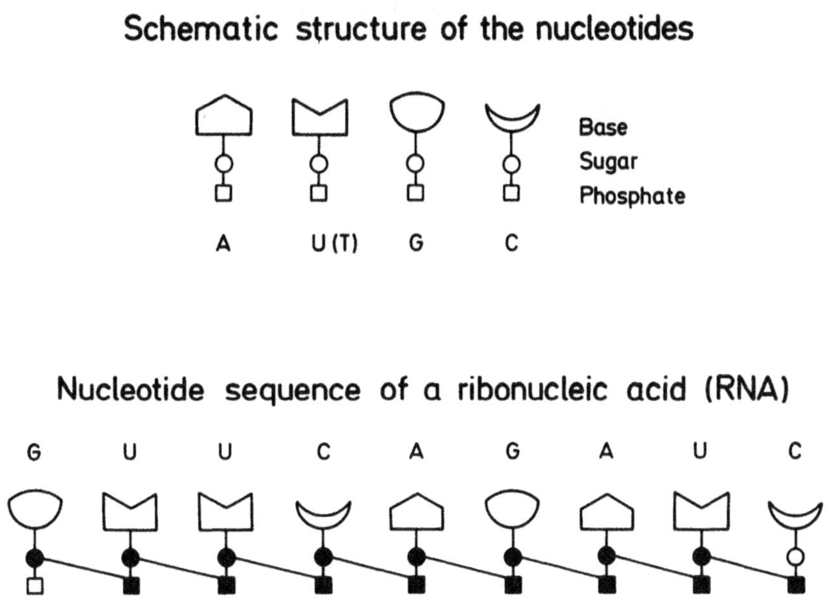

Fig. 2.1. Nucleic acid structure (schematic representation). All nucleic acids are built up from four chemical subunits (nucleotides). Each subunit is made up in turn of one organic base, one sugar and one phosphate group. The sugar can be ribose or deoxyribose, but only one of these forms occurs in any single nucleic acid molecule. There are accordingly only two classes of nucleic acid: ribonucleic acid (RNA) and deoxyribonucleic acid (DNA). A further difference is found in the fact that RNA contains the nucleotide U while DNA has the chemically very similar nucleotide T. The backbone of the polymer consists of alternate, covalently-linked sugar and phosphate units.

Fig. 2.2. Nucleic acid structure (chemical details). Each nucleotide has a specific and an unspecific region. The phosphate-sugar backbone is the same within each class (ribo- or deoxyribo-nucleotides) and the organic bases determine the identity of the nucleotide. Each phosphate group connects the 3′hydroxy group of one sugar to the 5′hydroxy group of the next. This imparts a specific directionality to a nucleic acid chain. The short chain depicted is RNA. In DNA the 2′ hydroxy groups are replaced by a hydrogen atom, and each H atom marked with an asterisk is replaced by a methyl (CH_3) group. The phosphate group is often depicted by an encircled P, as on the left-hand side. ▶

usually denoted by the initials of their chemical names. For ribonucleic acids (RNA) we have:

A (adenosine phosphate)

G (guanosine phosphate)

C (cytidine phosphate)

U (uridine phosphate)

whereas in deoxyribonucleic acids (DNA) the nucleotide U is replaced by the nucleotide

T (thymidine phosphate)

In the nucleic acid molecule the nucleotides are arranged - like the letters in a written language - in a long line. In this form they contain biological information. However, in order to understand a sequence of symbols unambiguously, one must first of all know in which direction they are to be read. In the case of the nucleic acids this direction is automatically signposted by the way in which the nucleotides are bound together chemically (Fig. 2.2). Further to this, a key is necessary for decoding the molecular message; we shall discuss this in detail in Section 2.3.

At this point we must consider how the biological information stored in the nucleic acid molecule is copied. This question is of course directly related to that of the molecular mechanism of inheritance.

Adenosine – 5'– phosphate

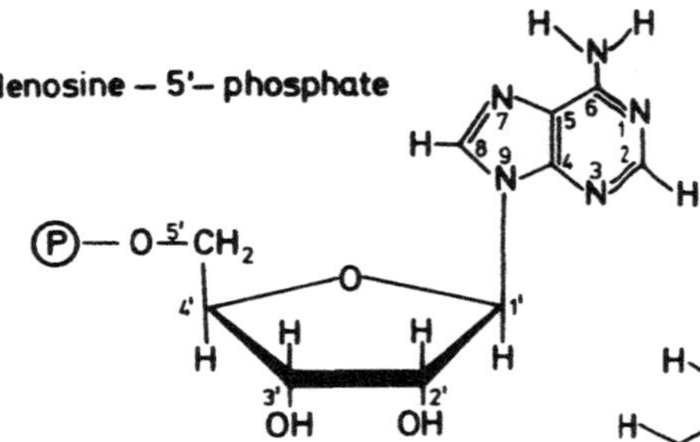

Cytidine – 5'– phosphate

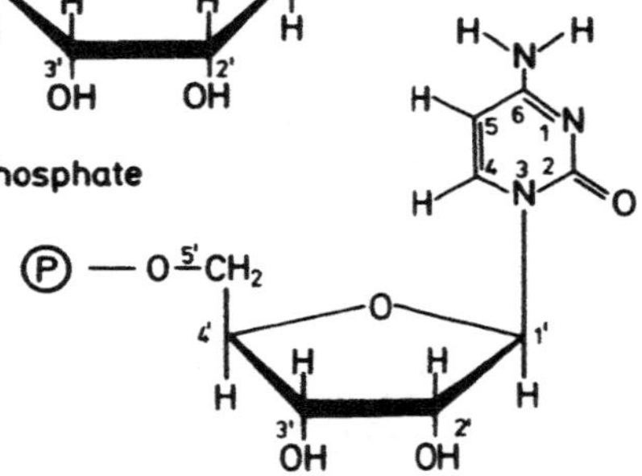

Guanosine – 5'– phosphate

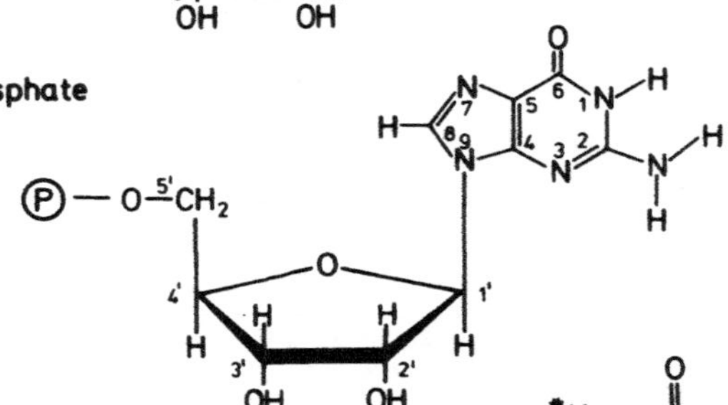

Uridine – 5'– phosphate

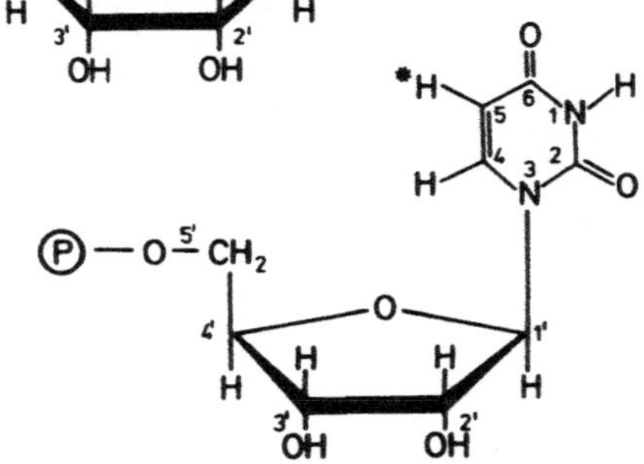

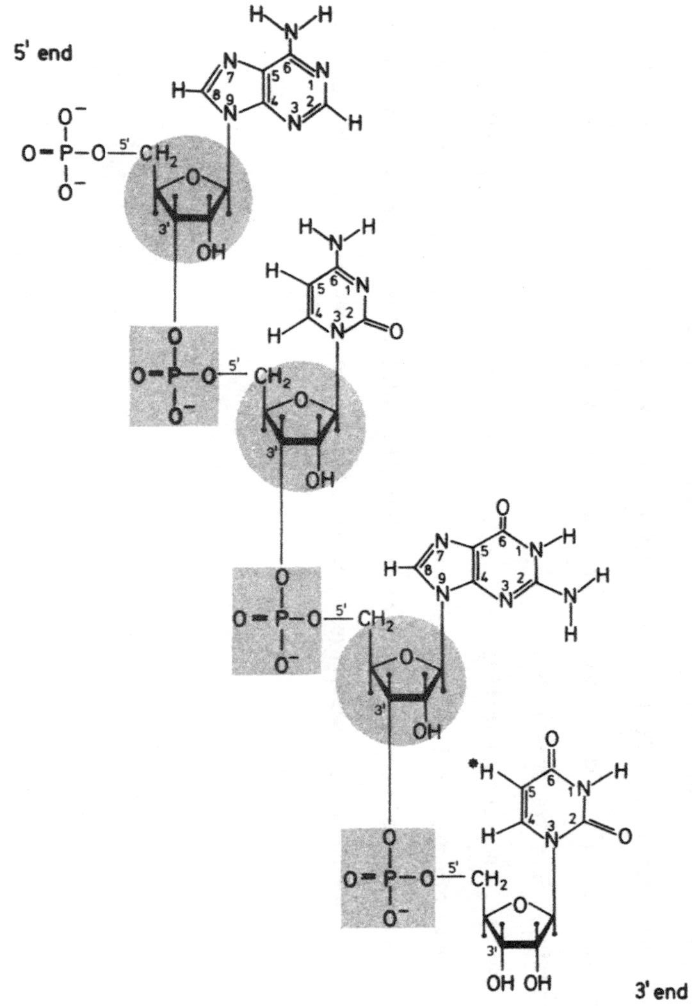

The ability of the nucleic acids to reproduce themselves is conferred upon them by their structure. To make this clear, let us examine these nucleotide building-blocks in more detail, and in particular the part which determines the nucleotide's identity, *viz.*, the organic base.

Two of the four nucleotide bases, cytosine and uracil (or thymine) belong to the category of bases called *pyrimidines*, whose basic chemical structure is a six-membered ring of carbon and nitrogen atoms (Fig. 2.3). The other two, adenine and guanine, are *purines*; these have an additional five-membered ring sharing an edge

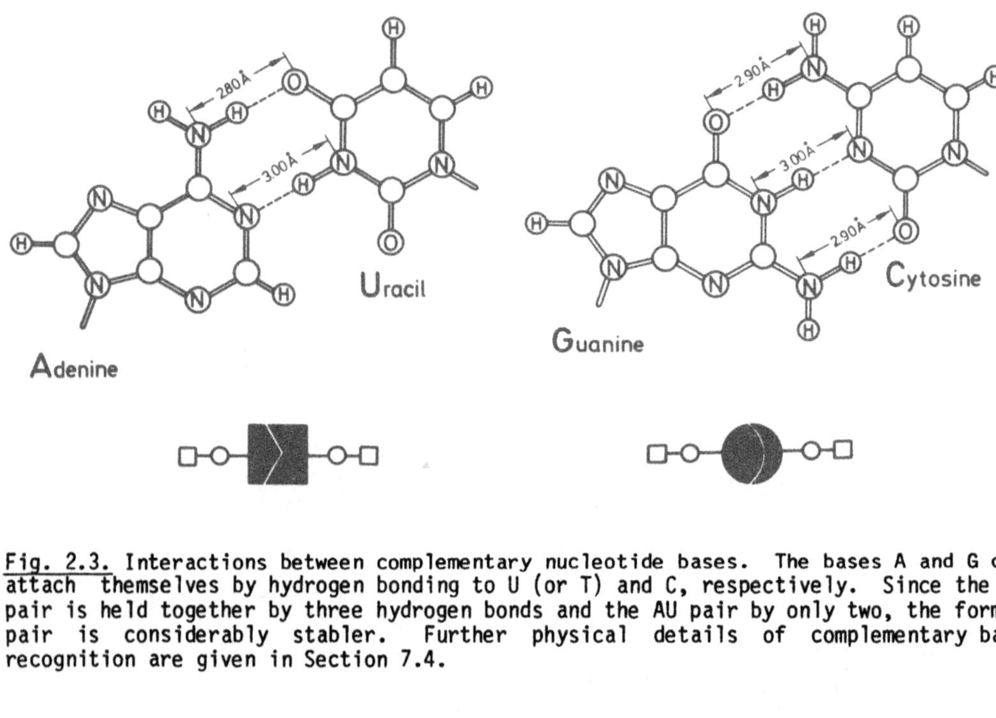

Fig. 2.3. Interactions between complementary nucleotide bases. The bases A and G can attach themselves by hydrogen bonding to U (or T) and C, respectively. Since the GC pair is held together by three hydrogen bonds and the AU pair by only two, the former pair is considerably stabler. Further physical details of complementary base recognition are given in Section 7.4.

with the pyrimidine ring. Specific *hydrogen bonds* can be formed between adenine and uracil (or thymine) on the one hand, and between guanine and cytosine on the other. The result is in each case a *base pair*, and the two kinds of base pair are almost exactly the same size and shape. In biochemical jargon: G and C are *complementary* to one another, as are U (or T) and A.

This selective interaction between the nucleotides provides the basis for the mechanism by means of which biological information can be passed on (Fig. 2.4). Thus a particular nucleotide sequence can accumulate a set of complementary nucleotides and make a "negative" copy of itself; in the same way the negative can be turned into a positive, so that the process of nucleic acid reproduction resembles the analogous procedure in photography. The positive and negative strands do not always separate after copying, as Fig. 2.4 suggests; more often they remain bound to each other in the well-known double helical structure, in which successive pairs of complementary bases are stacked flat upon one another.

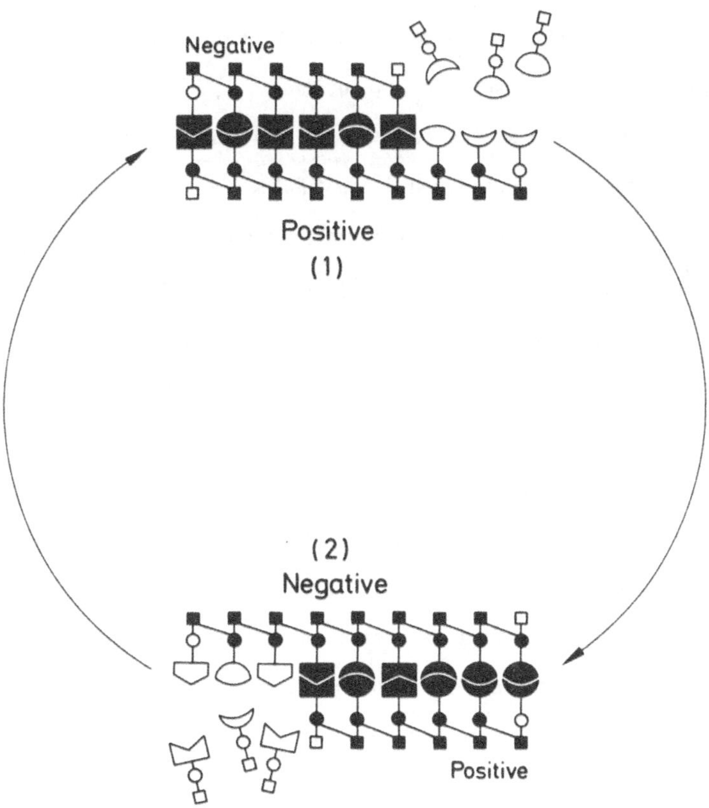

Fig. 2.4. Mechanism of nucleic acid reproduction. The reproduction of a nucleic acid has two principal phases: (1) synthesis of a negative form and (2) reversal of the negative form into a positive form. The copying of a single symbol involves complementary base recognition (see Fig. 2.3). To express the cyclical nature of this reproduction we shall frequently denote a self-reproducing nucleic acid molecule by the symbol .

The process of molecular reproduction rests principally upon the affinity between two complementary nucleotides. Because the energies of interaction are finite and the molecules are always in thermal motion, incorrectly-matched pairs occur with a finite probability (see Section 7.3), so that in an ensemble of nucleic acids undergoing reproduction a certain fraction of the copies will always be defective. Thus basic physical laws predict that the occurrence of mutations must be an inherent aspect of molecular genetics.

In the course of evolution Nature has developed special molecular copying-machines, the *polymerase* enzymes, which keep the error rate in nucleic acid reproduction relatively low and which, in addition, accelerate the copying process. Needless to say, a certain accuracy is needed in order not to lose the information accumulated over millions of years of evolution.

The relationship between the error rate and the greatest amount of information which can be reproducibly transmitted is of fundamental importance for evolution in general and for the origin of life in particular. The first to formulate this relation mathematically was M. Eigen (1971), and since then C. Weissmann and his co-workers have confirmed it experimentally (*cf.* Batschelet *et al.*, 1976; Domingo *et al.*, 1976). We shall discuss this error threshold relation in more detail in Chapter 7.

All our present knowledge of genetic information transfer at the molecular level indicates that the polymerases can keep the error rate in the incorporation of a nucleotide below a value of $3 \cdot 10^{-4}$. This sort of accuracy is just sufficient for the correct copying of the information stored in the smallest bacteriophages (the chain lengths of whose genomes are around 3500 nucleotides). In higher organisms sophisticated "proof-reading" mechanisms allow the copying of as many as 10^9 molecular symbols without loss of information. This is about the amount of information stored in a human genome.

The polymerases belong to the second large class of macromolecules which occur in organisms and which are of fundamental importance for them: the proteins. These are the chemical catalysts of living systems. They carry out synthesis, conversion of energy, regulation and other important biological tasks. The proteins are involved in all the processes of life, from metabolism to inheritance.

2.2. The Proteins

Proteins, like nucleic acids, are long-chained molecules. Their basic chemical units are the twenty common amino acids (Fig. 2.5). The "alphabet" of protein-language is thus much richer than the "alphabet" of nucleic-acid-language, which only has four "letters". As we shall now see, this has an important consequence for the structural variety of the proteins.

The step of greatest significance in the assembly of a protein is the formation of a *peptide bond* between the growing protein chain and each successive new amino acid

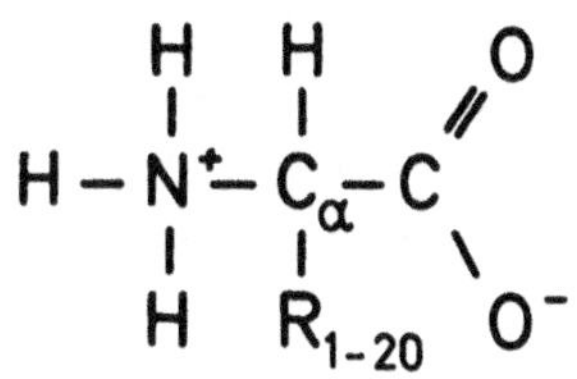

Fig. 2.5. Generalized structure of the amino acids. The C-atom in the α-position is substituted asymmetrically, *i.e.*, it connects four different atoms or groups. The single exception is glycine, in which the side-chain R is a hydrogen atom. Compounds with an asymmetric carbon atom always possess two possible configurations (L- and R-forms); each form is the mirror-image of the other. Interestingly, proteins found in organisms contain only L-amino acids.

(Fig. 2.6). The side-chains of the amino acids do not take part in this reaction, so that their individuality is still retained after polymerization.

In consequence of this a great variety of interactions can take place within each protein between the individual amino-acid constituents, and the peptide chain folds up into a characteristic, three-dimensional structure. Therefore an adequate description of the structure does not end with a statement of the linear sequence of amino acids (the *primary structure*) but requires further the concepts of a secondary and a tertiary structure: the *tertiary* structure is defined as the complete, three-dimensional structure of a single protein chain while the *secondary* structure refers to those parts of the molecule which possesses particular symmetry.*

The secondary structure of proteins arises through hydrogen bonds between the N-H (imino) groups and the C=O (carbonyl) groups of the protein chain. This can give rise to a corrugated planar structure (hydrogen bonds between parallel or

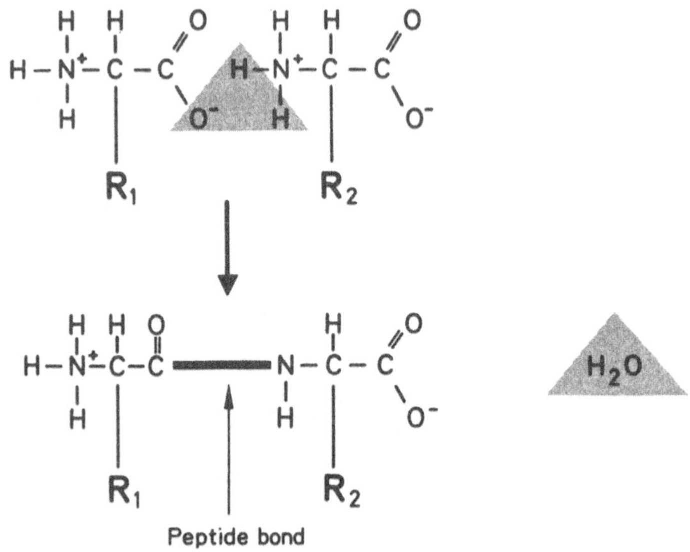

Fig. 2.6. Peptide bond between two amino acids. Two amino acids form a dipeptide when the amino group of one reacts with the carboxyl group of the other, eliminating water to give a peptide bond.

antiparallel segments of the peptide chain) or to the so-called α-helix (a cork-screw-like structure held together by hydrogen bonds between each C=O group and its third-nearest N-H group).

---

*The arrangement in space of several protein chains in a defined configuration is referred to as *quaternary* structure.

Highly symmetrical protein structures are almost unknown in Nature. There are several stereochemical reasons for this. One is that the various interactions between the side-chains of the amino acids lead to a more complex arrangement than a simple, regular secondary structure. Further, an α-helix is always interrupted whenever the "amino" acid proline appears. This is because proline is not really an amino acid at all: it has an imino (N-H) instead of the usual amino (NH_2) group and for this reason cannot join in hydrogen bonding to the third-nearest amino acid. Finally, when the side-chains of two cysteine residues come into contact, disulphide bridges (S-S) can form, and since these are covalent bonds they provide considerable stabilisation of the protein's tertiary structure.

The upshot of all this is that the tertiary structure of a protein is, under given boundary conditions, fully determined by the linear amino acid sequence and it appears spontaneously as a result of the physical interactions between its component amino acids.

The complexity and at the same time the order of the molecular structure which is produced by the manifold intramolecular interactions in a polypeptide chain is illustrated by the structural model of myoglobin (Fig. 2.7), a protein used in muscle to store oxygen. Myoglobin, along with haemoglobin, was one of the first biological macromolecules whose spatial structure was revealed by X-ray crystallographic analysis.

The structural chemistry of the proteins is naturally a good deal more complex than presented here. A detailed treatment would exceed our brief; of interest here are the *functional* consequences of protein structure.

The folds in the peptide chain bring together particular amino acids which in the protein sequence could lie wide apart. This can produce, out of two or more side-chains, a chemically functional unit with the ability to catalyse certain chemical reactions. Proteins with such catalytic properties are called enzymes; we have already made the acquaintance of one class of these, the polymerases. The functionally active side-chains make up the *active site* of the enzyme. Changes in the primary structure of an enzyme molecule often cause changes in the molecule's tertiary structure, which usually result in the loss of the enzyme's catalytic activity. Thus not only the tertiary structure of a protein but also its functional properties are determined by its primary structure.

We have already implied that each protein has its particular, specialized task within the cell. A complicated biochemical reaction is always helped along by several enzymes, of which each one catalyses a specific reaction step (*e.g.* the transfer of a chemical group from one molecule to another). However, enzymes are not

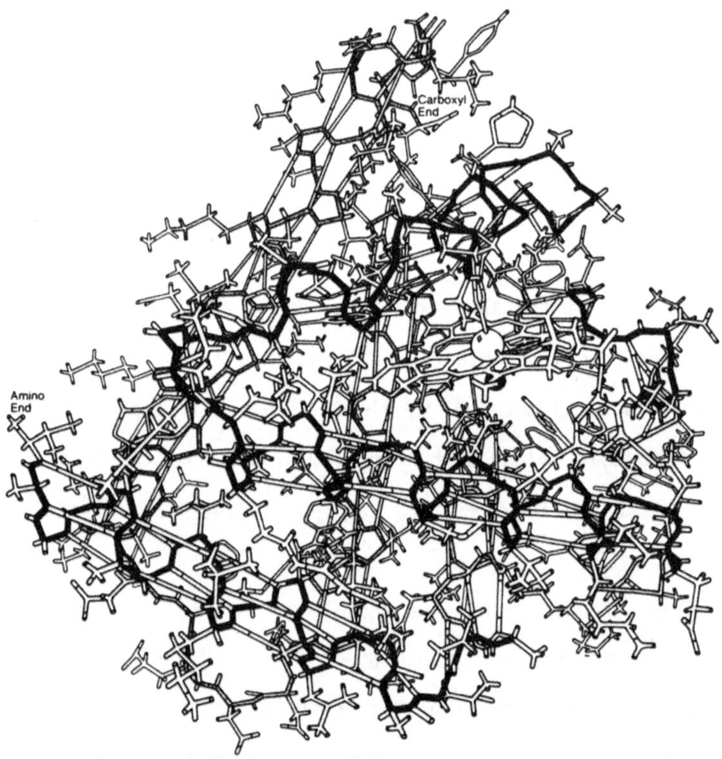

Fig. 2.7. Model of myoglobin. Myoglobin is a relatively small protein. It is made up of "only" 153 amino acids. Its spatial structure was elucidated, using X-ray structural analysis, by J.C. Kendrew and co-workers. The co-ordinates of the positions of each of the 2500 atoms in myoglobin can today be given to within ± 2 Å. The picture is based on a so-called ball-and-stick model, in which each atom is shrunk down to a point, leaving the covalent bonds visible as short "sticks". The course of the main chain is indicated by the dark line. The large ball represents the central iron atom, and the small one next to it shows the position of a water molecule, which is replaced by an oxygen molecule in oxygenated myoglobin. In spite of the complicated-looking arrangement of the atoms, there seems to be nothing disordered in the structure of myoglobin. (Illustration after Kendrew, 1961.)

only *reaction-specific*, but also *substrate-specific*, that is, specific in respect of the substance whose chemical transformation they catalyse.

Even in a simple organism like the bacterium *E. coli*, about a million protein molecules co-operate to keep the system reproducing itself. In doing this they carry out some two to three thousand different functions, which are kept in step with each other by delicate regulatory mechanisms.

At the beginning we showed how the nucleic acids make up the information store of living systems. We can now put this more precisely: *in the nucleic acid lies the blueprint for all the proteins in the living organism.* An example of this, the nucleic acid of the virus MS2, is shown in Figure 2.8. In viruses and in the bacterial cell these blueprints are all contained in a single nucleic acid molecule, the *genome*. The metaphor of the "skein of life" has here its concrete counterpart.

2.3. The Genetic Code

The code according to which the blueprint of a protein is encapsulated in a nucleic acid molecule has already been solved. Since the specification of the amino acid sequence of a protein is sufficient to determine the protein's structure (in a given physical environment) and thus its function, it only calls for a linear correspondence between the nucleotide sequence of a nucleic acid and the amino acid sequence of a protein for the information content of the blueprint to be defined in turn. It is obvious that a simple, one-to-one code scheme is insufficient, since protein-language has twenty symbols and nucleic-acid-language only four. Therefore the "vocabulary" of the nucleic acids is enlarged by the use of three-lettered code words, so that not 4 but 64 = (4^3) coding units are available. It is, incidentally, interesting that this logical requirement for unambiguous translation coincides with the physico-chemical fact that the triplet code represents an optimal compromise between the conflicting requirements of speed and accuracy of transmission.

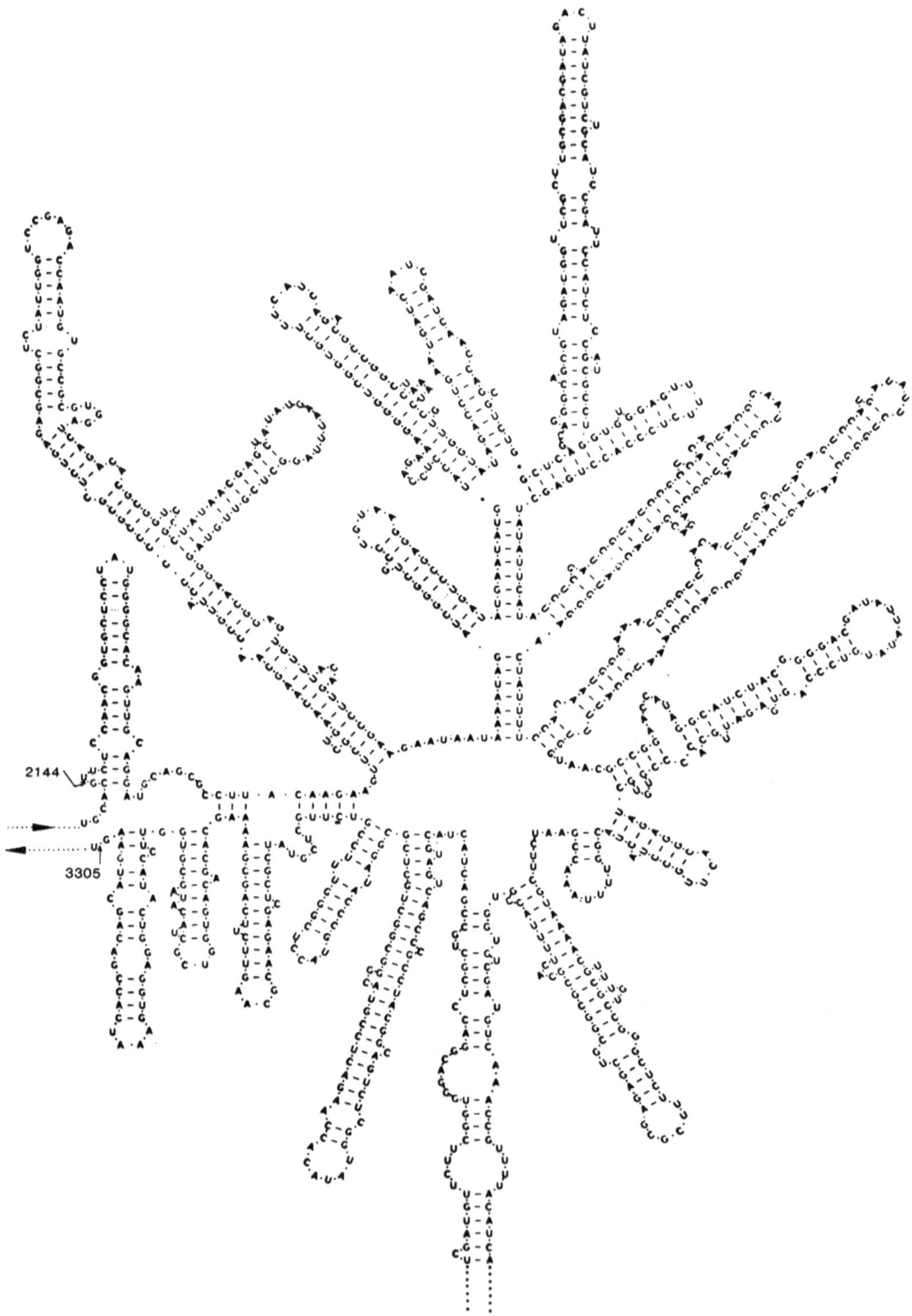
2144
3305

◄ Fig. 2.8. Excerpt from the genetic blueprint of bacteriophage MS2. The molecular alphabet of living beings is universal. It employs four molecular symbols, denoted here by the initial letters of their chemical names. A code-word consists of three letters. The exact sequence of code-words in the blueprint encodes the information needed to construct a molecular machine which serves in turn to stabilize and to reproduce the macromolecular blueprint. The sequence shown here encodes part of the replicase protein of bacteriophage MS2. With its 1632 nucleotides, this piece of nucleic acid represents about 30 per cent of the complete genome.

Physical interactions between the single nucleotides can fold up the blueprint, in a way such as shown here. The information content is however determined solely by the linear arrangement of the nucleotides. (After Fiers *et al.*, 1976.)

The distribution of rôles in the living cell between DNA, RNA and proteins can be regarded in the following way. The DNA represents the biological "headquarters" for information, carrying out two functions: it serves both as a self-reproducing store of information and as a template for transcription. The scheme according to which genetic information in the cell is processed, with the help of RNA and proteins, is fairly well understood. At the beginning the biological information is copied (transcribed) from the DNA molecule into a messenger RNA (mRNA) molecule. The mRNA then transports the genetic information to the ribosomes. Ribosomes are complex functional units composed of ribonucleic acid and protein molecules; they perform the actual biosynthesis of proteins. Thus their basic task is to decipher the information contained in the nucleotide sequence of the mRNA and to turn it into a polypeptide containing the corresponding amino acid sequence. In doing so it makes use of adaptor molecules, which are equipped both with an *anticodon* complementary to the codon which it is to read and with the corresponding amino acid (aminoacyl tRNA). The adaptor molecules bind two at a time to two consecutive codons on the mRNA, and the growing peptide chain is transferred, a step at a time, from one tRNA molecule to the next, thus growing by one amino acid at each transfer.

The individual assignment of nucleotide triplets to particular amino acids, the so-called *genetic code*, is shown in the box in Fig. 2.9. Because there are 64 triplets and only 20 amino acids, the genetic code is redundant. In some cases up to six different triplets code for the same amino acid.

The degeneracy of the genetic code in respect of the third position in the codon (the 3′ end) may be explained by the "wobble" hypothesis of F.H.C. Crick (1966).

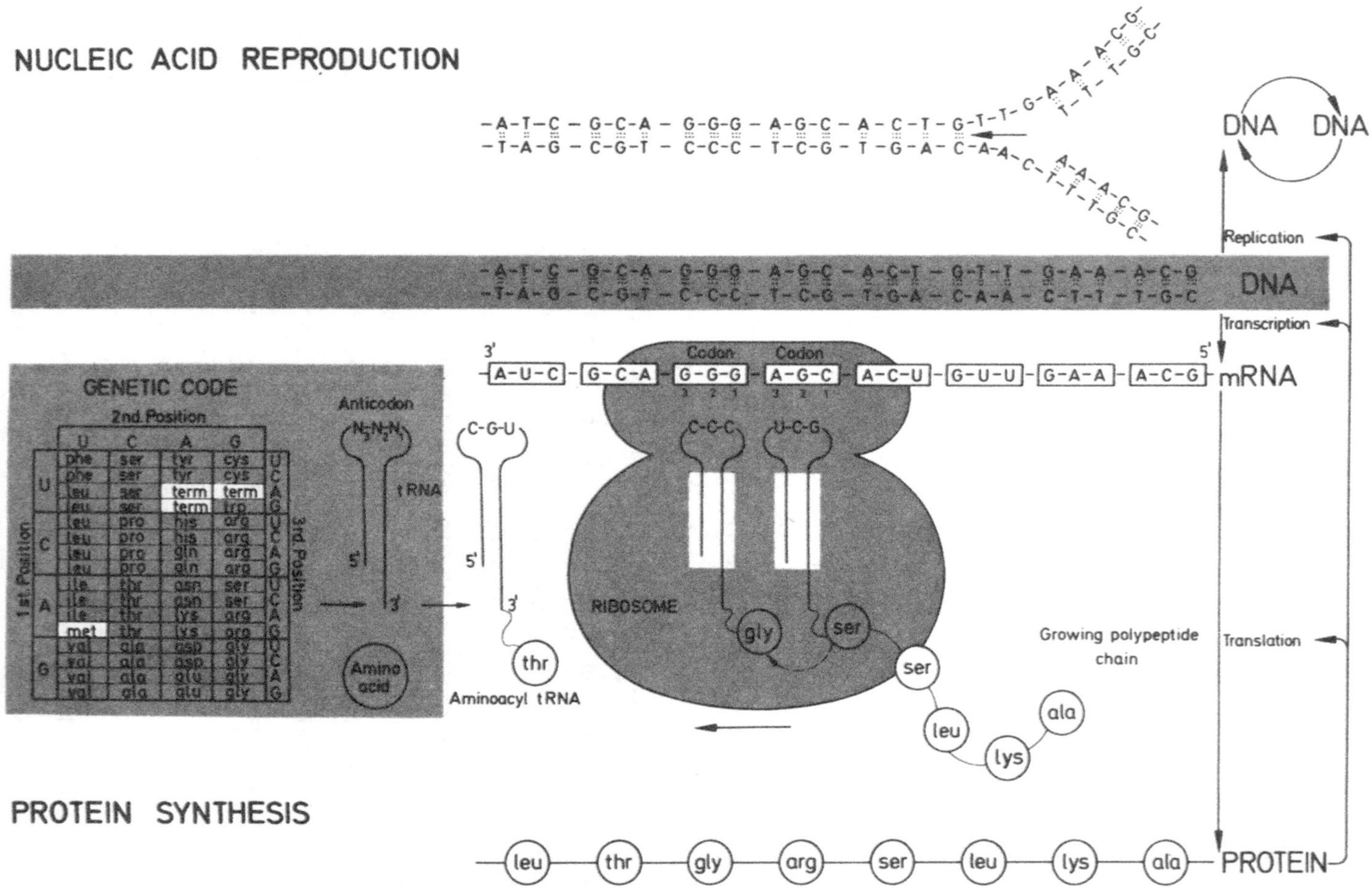
NUCLEIC ACID REPRODUCTION
PROTEIN SYNTHESIS
DNA
DNA
Replication
Transcription
Translation
mRNA
PROTEIN
3'
5'
A-U-C
G-C-A
G-G-G
A-G-C
A-C-U
G-U-U
G-A-A
A-C-G
Codon
Codon
C-C-C
U-C-G
C-G-U
RIBOSOME
Growing polypeptide chain
Aminoacyl tRNA
Anticodon
tRNA
Amino acid
GENETIC CODE
2nd. Position
1st. Position
3rd. Position
U C A G
U: phe ser tyr cys U; phe ser tyr cys C; leu ser term term A; leu ser term trp G
C: leu pro his arg U; leu pro his arg C; leu pro gln arg A; leu pro gln arg G
A: ile thr asn ser U; ile thr asn ser C; ile thr lys arg A; met thr lys arg G
G: val ala asp gly U; val ala asp gly C; val ala glu gly A; val ala glu gly G
leu thr gly arg ser leu lys ala

◄ Fig. 2.9. Simplified scheme of nucleic acid reproduction and protein biosynthesis. All organisms - be they viruses, procaryotes or eucaryotes - employ a similar genetic apparatus and the same genetic code for the translation of nucleic acids into proteins. This code contains not only nucleotide triplets which function as genetic words but also ones which function as punctuation marks. Thus the codon AUG, at the beginning of the cistron, means "start protein synthesis". There are also three "stop" codons. The abbreviated names for the amino acids are:

ala = alanine	gly = glycine	pro = proline
arg = arginine	his = histidine	ser = serine
asn = asparagine	ile = *iso*leucine	thr = threonine
asp = aspartic acid	leu = leucine	trp = tryptophan
cys = cysteine	lys = lysine	tyr = tyrosine
gln = glutamine	met = methionine	val = valine
glu = glutamic acid	phe = phenylalanine	

Each amino acid is carried by a specific adaptor, called transfer RNA (tRNA), which further possesses the corresponding anticodon. The adaptor molecule brings the activated amino acid into the correct position on the messenger RNA (mRNA). Thus the protein is assembled, step by step, on the ribosome.

This postulates that the nucleotide G can form a reasonably stable base pair with the nucleotide U as well as with its normal partner C. This leads to interactions such as the following (N_+, N_- are undefined complementary nucleotides):

Codon (mRNA)	5′....... $N_+N_+{}^{C}_{U}$ ....3′		5′....... $N_+N_+{}^{A}_{G}$ ....3′
	⋮ ⋮ ⋮	and	⋮ ⋮ ⋮
Anticodon (tRNA)	3′....... N_-N_-G ....5′		3′....... N_-N_-U ....5′

According to the "wobble" hypothesis, A and G or C and U can be regarded as equivalent in the first position of the anticodon.

Translation of a nucleotide sequence into the corresponding amino acid sequence does not take place spontaneously. It requires the catalytic assistance of many proteins. The essential steps of protein biosynthesis are shown in Fig. 2.9, again in greatly simplified form.

The whole apparatus of protein biosynthesis is thus a highly-organized, self-regulating system, in which proteins give rise to nucleic acids and *vice-versa*. According to what we know at present, both the logical and the material components of this system are uniform throughout the living world. There also seems to be no organism - be it a virus, a bacterium or a human being - which does not use the same genetic code*. Thus the results of molecular genetics provide spectacular corroboration of the thesis of Charles Darwin (1859) that all living beings are descended from a common ancestor.

---

*This statement holds for all *autonomous* organisms. However, in protein biosynthesis within mitochondria deviations from the common code scheme have recently been observed (*cf.* Hall, 1979; Bonitz *et al.*, 1980).

3. The Information Problem

The molecular description of living systems has given us some insight into the roots of biological complexity. As a starting-point for our quantitative consideration we shall now attempt to grasp the extent of this complexity.

The smallest living organisms with their own metabolism are the bacteria. Even these need almost four million nucleotides in order to encode their genetic information in a nucleic acid molecule. The information capacity present in so many symbols can be compared, by way of illustration, to that of human language: four million nucleotide letters would fill a 1000-page book.

Just as changing one or more letters can distort the meaning of a written text, a mutation in the microstructure of a gene can have far-reaching macroscopic consequences, such as the death of the organism carrying the mutation. The exact sequence of symbols in the bacterium *E. coli* is however only one of $10^{1\,500\,000}$ combinatorially possible sequences.

This impressive numerical example leads to the conclusion of J. Monod (1972) that the genetic blueprint of even the simplest organism is as unlikely to have arisen by chance as a comprehensive text-book of biology is likely to have arisen by shaking up all its letters. Since however organisms do exist on Earth, Monod saw in the origin of life a singular event, which, like a big lottery win, represents an arbitrarily improbable and therefore, for the winner, unique event. If it were indeed correct to regard the origin of life as the result of "blind" chance, then Monod's conclusion would have far-reaching philosophical implications. However, we shall see that Monod's hypothesis of chance is only *one* aspect of evolutionary reality. The complementary aspect of natural law and necessity was - for whatever reason - neglected by Monod.

It is an epistemological foundation of modern science that it endeavours, by formulating scientific laws, to reduce as far as possible the rôle of chance in the explanation of phenomena. We shall now attempt to show that this is to some extent possible for the question of the origin of life. Let us illustrate this with a concrete example, again taken from the world of biological macromolecules.

The analysis of living systems has shown that even the smallest catalytically active proteins are built up from a hundred amino acids or more. We also know that the physiological function of a protein depends upon the order of its constituent amino acids, *i.e.*, its primary structure. The number of permutations which is needed to cover all possible sequences 100 amino acids long amounts to

$$N = 20^{100} \approx 10^{130} \tag{3.1}$$

Even if a point mutation occurs every 10^{-6} seconds, it would still take a time

$$t \approx 10^{124}\ \mathrm{s} \tag{3.2}$$

to cover all possible sequences (*cf.* Fig. 3.1). But the age of the world is "only"

$$t \approx 10^{17}\ \mathrm{s} \tag{3.3}$$

The probability of reaching by pure chance a defined protein structure, optimally adapted to some purpose, is thus, within existing dimensions of time and space, as good as zero.

We can instead - at least in a thought experiment - proceed systematically. Here we test each of the 100 amino acids individually for its functional quality and in each case keep the best result. In a first approximation every twentieth point mutation will be successful, so that the optimal structure is reached after only

$$N = 100 \cdot 20 \approx 10^{3.3} \tag{3.4}$$

trials. In reality, the optimization is a co-operative process, in which the optimal occupation of one position depends upon the properties of the occupants of other positions. This is particularly true of the assembly of the active centre of

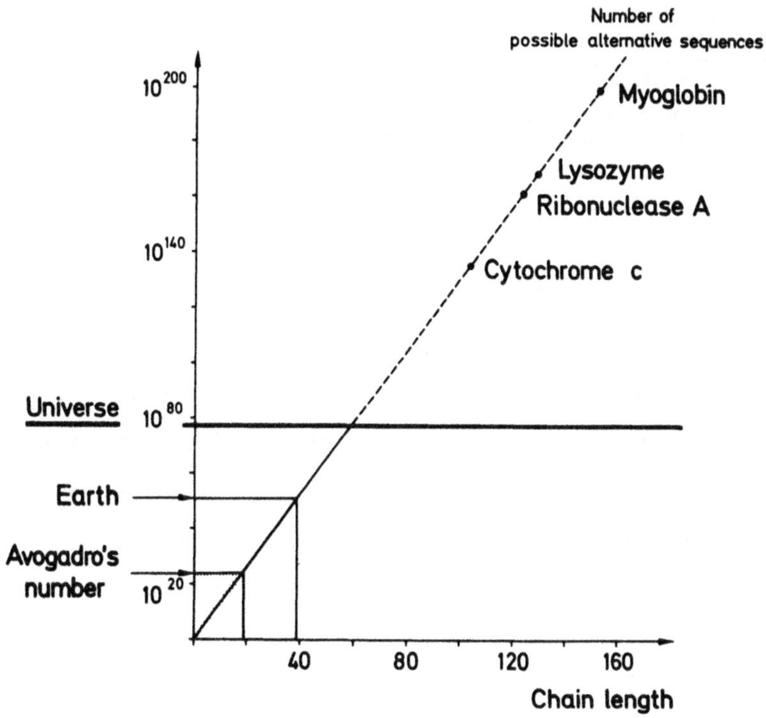

Fig. 3.1. The complexity of catalytically active protein structures. The electron-transport enzyme cytochrome c, with 104 amino acids in its chain, is one of the smallest naturally-occurring proteins with catalytic function. Even so, a protein of this length has 10^{135} possible alternative sequences. In comparison, the mass of the universe is 10^{78} times that of a hydrogen atom, and that of the earth a mere 10^{51} times this unit.

an enzyme molecule. If two, three, or more positions act co-operatively then the probability of finding the ideal pair, triplet *etc.* is only $(1/20)^2$, $(1/20)^3$,... . None the less, the selective method remains greatly superior to that of pure chance.

At this point we should perhaps re-examine these statements more critically. Up to now we have assumed implicitly that only *one*, if any, of the 10^{130} alternative sequences for our minimal protein structure has a biological function. This is certainly incorrect. Numerous sequence analyses of biologically active proteins have

Fig. 3.2. Phylogenetic family tree from comparative sequence analysis of the protein cytochrome c (*cf.* Fitch and Margoliash, 1967). The numbers on the branches are the minimum numbers of nucleotide substitutions in the DNA of the genes that could have given rise to the observed differences in the amino acid sequences. The overall relations agree fairly well with those inferred from the fossil record and other sources. There are only a few deviations of the cytochrome c phylogeny from the traditional one. Thus the chicken appears to be related more closely to the penguin than to ducks and pigeons; the turtle, a reptile, appears to be related more closely to birds than to the rattlesnake, and man and monkeys diverge from the mammals before the marsupial kangaroo separates from the placental mammals. Despite these differences, the amino acid sequence of cytochrome c reflects astonishingly well the evolutionary kinship of organisms as observed on the macroscopic level. (From Ayala, 1978.) ▶

revealed that as a rule many mutants of one protein have the same function. Substantial loss of functional activity occurs only when the mutation alters or destroys the structure of the active centre. A frequently-cited example of the variability of the primary structure of a protein, keeping an invariant function, is that of cytochrome c. This molecule shows variations in its primary structure from one species to another, and these have been utilized to construct a macroscopic tree of evolution (Fig. 3.2). This shows clearly that biologically meaningful function can be fulfilled by more than one out of the 10^{130} possible variations of a protein sequence. This assumption lies at the bottom of the selective optimization process which we have discussed as an alternative to the chance hypothesis.

We shall probably never find out how much freedom a biologically active protein has for variation in its primary sequence. To do this would require in principle the testing of all the 10^{130} permutations - an impossible task, as shown above. This means, strictly speaking, that we cannot rule out the correctness of Monod's hypothesis of chance. Our only argument against it is that it is *unscientific*: a particular characteristic of science is its attempt to eliminate the rôle of chance, as far as possible, by propounding explanations based upon regular behaviour. The discomfort caused by a persistent residue of chance has already been pointed up by the stormy discussions around the interpretation of quantum mechanics.

We shall therefore avoid explaining the origin of life as a singular, random event, and shall attempt to show that the origin of life is an inevitable process if certain prerequisites are fulfilled. At this point we must eliminate a possible mis-

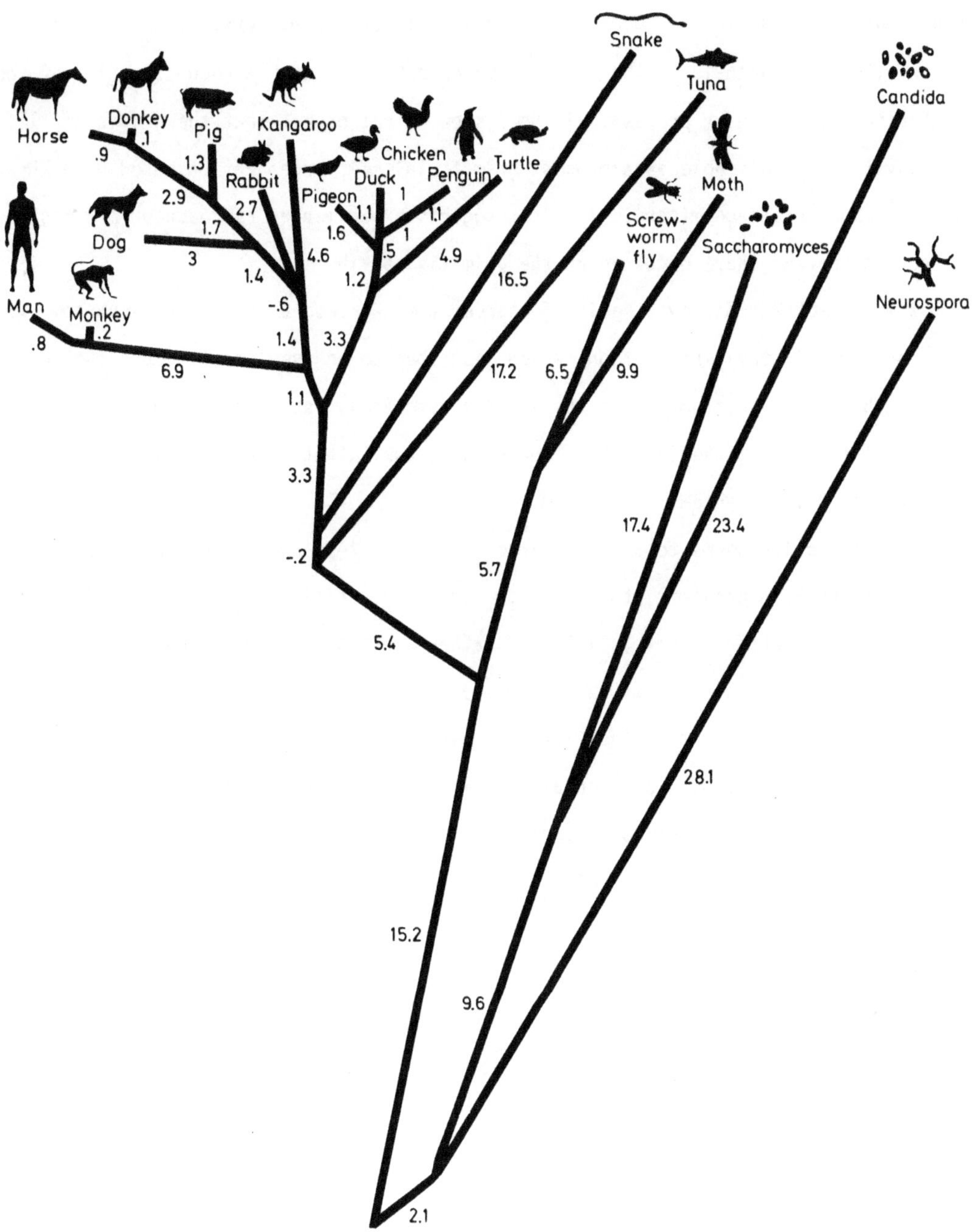
Snake
Tuna
Candida
Horse
Donkey
Pig
Kangaroo
Chicken
Turtle
Rabbit
Duck
Penguin
Pigeon
Moth
Screw-worm fly
Saccharomyces
Dog
Man
Monkey
Neurospora
.9
.1
1.3
2.9
2.7
1.7
3
1.4
-.6
4.6
1.6
1.1
1
1.1
1
.5
4.9
1.2
16.5
.8
.2
1.4
3.3
6.9
17.2
6.5
9.9
1.1
3.3
-.2
17.4
23.4
5.7
5.4
28.1
15.2
9.6
2.1

understanding. The theory to be developed is a theory *a posteriori*. It does not allow either a prognosis of the future development of life or a reconstruction of the historical evolutionary process. It does show in retrospect that the origin of life is physically explicable as a process of material self-organization based upon known laws of physics and chemistry. It also suggests experimental approaches for testing processes that *may* have occurred on the primitive earth.

If we reject Monod's hypothesis of chance, then we need a selection mechanism for the origin of the information which is then to be used for the construction of complex biological systems. But is it not precisely such selection - as many representatives of the holistic biology will ask - that is a typical and exclusive property of living systems?

This objection seems to steer us into a mental *cul-de-sac*, unless we choose after all to revert to a creation story or to an existentialistic concept of chance. Avoiding the latter course, let us rather re-examine the thesis that natural selection operates only on *living* systems by looking at the conditions required to produce selection at the (non-living) molecular level. The primary question will be: "What properties must a material system possess in order to organize itself, spontaneously and making use of selection?"

II. Principles of Molecular Selection and Evolution

4. A Model System for Molecular Self-Organization

A theory of evolution intended to describe the selective optimization of biological macromolecules will in the nature of things be founded on the laws of chemical reaction kinetics. Since we are going to want to formulate mathematically the principles of molecular selection and evolution, we shall need to concentrate our attention on a defined model system (*cf.* Chapter 13).

Consider the system shown in Fig. 4.1. It comprises basically a reaction vessel, in which biological macromolecules are continually built up out of energy-rich monomers. But thermal molecular movement curtails the lifetime of all the molecules and of all the ordered states that result from their interactions with one another. The biological macromolecules therefore decay, after a certain time, back to their (energy-deficient) molecular constituents.

We wish to prevent the reaction system from reverting into a state of thermodynamic equilibrium. We therefore assume that the reaction vessel has semi-permeable walls, through which the system perpetually exchanges energy and also matter with its surroundings: through one wall energy-rich monomers are introduced, while through the other the products of decay (energy-deficient monomers *etc.*) are removed. Defined reaction conditions can be set up by regulating the fluxes of energy-rich and energy-deficient monomers and the dilution flux Φ_0.

Let us now further specify the chemical components of our reaction system. A single macromolecule consists of ν molecular subunits, of which in turn μ different sorts may exist. Among biological macromolecules, μ is 4 for nucleic acids and 20 for proteins. The polymers can be divided according to length into ν classes. A measure of the information capacity of the polymer class ν is then given by

$$N = \mu^{\nu} \tag{4.1}$$

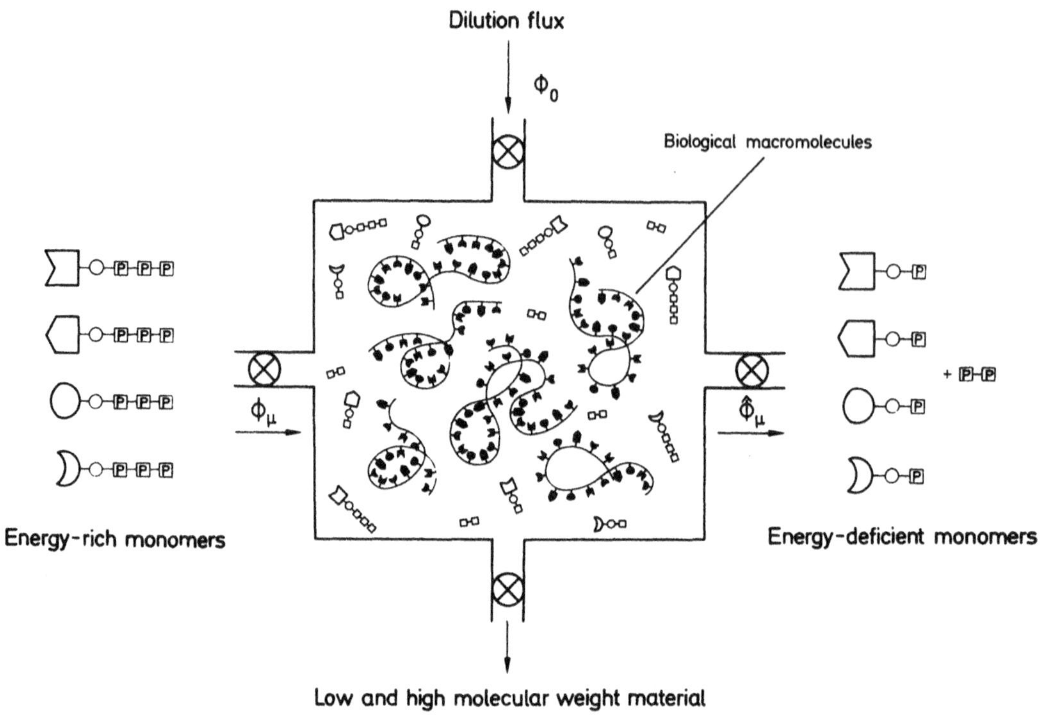

Fig. 4.1. Flow reactor to be used as a model system for molecular self-organization processes. In the reaction vessel biological macromolecules (*e.g.* nucleic acids) are continually built up from energy-rich monomers. Defined reaction conditions can be set up in the system by regulating the supply of energy-rich monomers $\boldsymbol{\Phi} = (\Phi_1,\ldots,\Phi_\mu)$ and a global dilution flux Φ_0, which effectively controls the total population of polymers. Thus the system can work, for example, at constant flux or at constant concentration.

The model system shown here can induce, under appropriate conditions, selection competition between the various self-reproducing polymers and can thus simulate basic processes of molecular evolution (see also Fig. 13.9).

where N is the number of all combinatorially possible sequences of ν symbols. In the following discussion we shall consider only polymers with a uniform chain length ν, and then show that while this assumption greatly simplifies the mathematical treatment, it does not restrict the validity of our conclusions.

Let us suppose that in the reaction system every possible sequence of length ν can in principle be present; there are

$$z_i \geq 0 \tag{4.2}$$

copies of the ith sequence. The total population of polymers of chain length ν is then given by the number

$$Z = \sum_{i=1}^{N} z_i \tag{4.3}$$

In the following treatment we shall frequently replace the population numbers z_i and Z by concentration terms x_i and c, so that

$$x_i = \frac{z_i}{V} \quad \text{and} \quad c = \frac{Z}{V} = \sum_i x_i \tag{4.4}$$

where V is the volume of the flow reactor.

It is easily apparent (*cf.* also Chapter 3) that it is only meaningful to consider self-organization mechanisms if for the particular class of polymers

$$Z \ll N \tag{4.5}$$

is fulfilled, *i.e.*, the expectation value of a particular polymer sequence is vanishingly small. Only then can we exclude the possibility that a distribution of all possible sequences already exists from the beginning in the system.

As we have indicated in Chapter 3, the relation (4.5) is always fulfilled for information carriers whose chain lengths are large enough to encode biologically useful functions.

Let us demonstrate this with a numerical example. A 1 millimolar solution of nucleic acids is for primordial conditions certainly an unusually concentrated solution. In a litre of this solution there are some 10^{21} nucleic acid molecules. Even in such a concentrated solution, the inequality (4.5) would be well satisfied for a nucleic acid chain of 100 nucleotides ($10^{21} \ll 4^{100} \approx 10^{60}$). However, 100 nucleotides would not even suffice to encode the information needed to build a self-reproducing genetic apparatus.

5. Deterministic Theory of Selection

We have emphasised more than once that the origin of biological information is above all a statistical problem: the principal question is how molecular self-organization can effect the choice of a particular polymer sequence out of an almost limitless number of alternatives.

The statistics of selection needed to solve this problem will first of all be formulated as a deterministic theory. That is, we shall initially disregard the influence of fluctuations, and shall assume that each polymer sequence is present in a sufficiently large number of copies. This deterministic approach enables us to neglect in particular the fluctuations in concentration which are ultimately due to quantum-mechanical uncertainty and which are therefore always present.

5.1. Phenomenological Equations

Let us first set up the kinetic equations for the reaction system in Fig. 4.1. For clarity we shall start by considering only the nucleic acids. The term *molecular information carrier* or *molecular species* will in the following treatment always denote a nucleic acid of a particular sequence. We should note at this point that our use of the term "species" is thereby different from that of the biologist; a biological species corresponds more closely with what we shall later denote a "quasi-species" (see Section 5.4).

M. Eigen (1976) has shown using simple game models that selection in the Darwinian sense is always based on certain dynamic properties (such as the rate of reproduction or decay) of the individuals (marbles, molecules *etc.*) participating in the process. In our case the nucleic acids are the participants in the selection process, and we must therefore start out from the rate equations which describe the reproduction and the decay of nucleic acids.

The rate of change in the concentration of the molecular species I_i is clearly given by the difference between the rates of processes leading to increase and decrease in this concentration. Denoting these by R_i and r_i respectively, we have

$$\dot{x}_i = R_i - r_i \qquad (\dot{x}_i \equiv dx_i/dt) \tag{5.1}$$

If we take into account the fact that the reaction vessel in Fig. 4.1 has semipermeable walls, we can distinguish between the rates of flow into and out of the vessel, R^* and r^*, and the rates of formation and decay, $\overline{R}$ and $\overline{r}$, within it. Thus

$$\dot{x}_i = (\overline{R}_i + R_i^*) - (\overline{r}_i + r_i^*) \tag{5.2}$$

Without going into details of the reaction kinetics, we can further specify the various rate terms on the basis of our model system, as follows.

(1) AUTOCATALYTIC FORMATION OF MOLECULAR INFORMATION CARRIERS. This we shall express as

$$\overline{R}_i = F_i x_i + \sum_{j \neq i} \psi_{ij} x_j \tag{5.3}$$

The first term on the right of Eq. (5.3) denotes the inherent, self-instructed reproduction of a nucleic acid I_i, represented by a general rate parameter F_i. This parameter may in turn be a function of the concentrations of energy-rich monomers $(m_1 \ldots\ldots m_\mu)$ or of other polymers (x_k). F_i may also be an explicit function of time.

The second term takes account of the fact that for fundamental reasons (outlined in Sections 2.1 and 7.3) the process of reproduction is never quite free of error, so that each species I_j has a certain probability ψ_{ij} of contributing to the population of species I_i.

It will later prove useful if we split F_i into two further, purely phenomenological terms:

$$F_i = A_i Q_i \tag{5.4}$$

The parameter A_i is a rate constant which tells us how fast new nucleic acids are synthesized on the nucleic acid template I_i, independently of whether the copies are correct or not. We can call A_i an *amplification factor*.

The accuracy of the synthesis is given by a *quality factor* Q_i, which tells us with what probability the synthesis of I_i on the matrix will be error-free. Thus Q_i always lies within the limits

$$0 \leq Q_i \leq 1 \tag{5.5}$$

and $1-Q_i$ is the probability of erroneous copying of species I_i.

(2) INFLUX OF MOLECULAR INFORMATION CARRIERS. In our model system no nucleic acids flow into the reaction vessel from outside, so for each species I_i

$$R_i^* \equiv 0 \tag{5.6}$$

(3) DECAY OF MOLECULAR INFORMATION CARRIERS. The decay of the molecular species I_i by chemical decomposition will be a reaction of at least first order, so that

$$\overline{r}_i = D_i x_i \tag{5.7}$$

D_i is a general rate parameter which itself may depend on the value of any of the concentration variables x_k.

(4) EFFLUX OF MOLECULAR INFORMATION CARRIERS. A regulable, global dilution flux Φ_0 can be used to allow the total polymer concentration to reach any desired value at any time. Since this dilution flux does not distinguish between individual information carriers, the rate of dilution r_i^* is proportional to the concentration of species I_i, *i.e.*

$$r_i^* = \varphi x_i \qquad \text{with} \qquad \Phi_0 = \sum_i r_i^* = \varphi \sum_i x_i \tag{5.8}$$

where φ is a general constant of proportionality.

As long as the spontaneous and non-instructed formation of the molecular species I_i can be neglected, we can easily deduce the fundamental conservation equation

$$\sum_i A_i(1-Q_i)x_i = \sum_i \sum_{j \neq i} \psi_{ij} x_j \tag{5.9}$$

This is the statement that all errors in the reproduction of a species make a contribution to the production of other species, so that the total rate of production of mutants in our system is the same whether we count the total losses due to poor copying of all species (left-hand side) or the contributions which these make to the populations of other species (right-hand side).

Substitution of Eqs. (5.3) to (5.8) into Eq. (5.2) leads to the mass-balance equation for each molecular species I_i

$$\dot{x}_i = (A_iQ_i - D_i - \varphi)x_i + \sum_{j \neq i} \psi_{ij} x_j \qquad (i,j = 1,\dots,N) \tag{5.10}$$

This set of phenomenological differential equations describes in general terms the population dynamics of molecular information carriers which are kinetically independent of each other and which possess the following three properties:

(a) *Metabolism*, expressed in the two rate terms $\sum_i A_i x_i$ and $\sum_i D_i x_i$, which describe the turnover of energy-rich monomers into energy-deficient ones. Since matter and energy can flow in and out in the form of monomers, the system is thermodynamically open, a point which we shall develop in Section 12.1.

(b) *Self-reproduction*, expressed in the structure of the rate equation (5.3), in which the formation rate is proportional to the concentration x_i, independently of the way in which F_i depends on x_k.

(c) *Mutability*, expressed by the quality factor Q_i, which in real systems is always less than unity.

In Chapter 1 we stated that these three properties are common to all living systems.

It can be shown that the kinetic equations (5.10) do not yet lead to competitive selection in our model system. To see this we must integrate these equations, which on account of the non-linearities involved is not an easy task. However, the principal properties of the solutions can be examined using simplifying approximations.

For this purpose we eliminate the non-linearities with the help of the assumption that all A_i- and D_i-parameters are independent of the concentration variables x_k. Furthermore, we assume that the A_i- and D_i-parameters are constant in time and that all ψ_{ij} are negligibly small (*i.e.* $Q_i \approx 1$). Under these assumptions the solutions of Eq. (5.10) are

$$x_i(t) = x_i(0)\exp\{(A_i - D_i - \varphi)t\} \qquad (i = 1,\ldots,N) \tag{5.11}$$

Thus there is a threshold relationship

$$A_i \gtrless D_i + \varphi \tag{5.12}$$

which in our model leads to the phenomenon of *segregation*: all species with

$A_i > D_i + \varphi$ increase in number as time goes on, and all those with $A_i < D_i + \varphi$ die out. In this case there is no competition and thus no selection.

In order to induce *selection pressure* in a system such as that described by Eq. (5.10), we must subject it in some way to limitation of overall growth. A direct way to limit growth in our model system of Fig. 4.1 is to subject it to dilution using the dilution flux Φ_0. We can, for example, regulate the dilution flux in step with the total polymer population (*i.e.*, concentration, as the volume is fixed), so as to keep the latter constant in time. We can then speak of a system with *constant population* (CP). A CP system thus satisfies the relation

$$c = \sum_i x_i = \text{const.} \qquad (5.13)$$

This kind of limitation of overall growth is the most convenient for our purposes, as we shall see in Part IV. This naturally does not mean that the historical course of evolution ever really took place under such idealized conditions. On the contrary, it seems more likely that a limited supply of energy (*e.g.*, finite sources of nourishment), or a continually changing combination of the limitations upon growth already mentioned, was responsible for the historical causes of selection pressure. We shall return to these questions in Section 12.2. The selection pressure induced by dilution is in this case non-specific. Dilution affects the entire population and does not discriminate among different molecular species.

As an alternative to the selection pressure of constant population we shall discuss a different kind of limit placed upon growth in which the total concentration of the polymers is allowed to vary, while the inflow of the energy-rich monomers $(\Phi_1, \ldots, \Phi_\mu)$ is invariant.
A system with *constant flux* (CF) thus satisfies the condition

$$\Phi_1, \ldots, \Phi_\mu = \text{const.} \qquad (5.14)$$

The constraints CP and CF are connected respectively with the thermodynamic boundary conditions of constant driving forces of reaction (that is, constant thermodynamic affinity) and of constant reaction flux. Both lead to stationary states. These conditions are connected with the near-steady-state reaction system discussed by P. Glansdorff and I. Prigogine (1971); we shall explain them in Section 12.1.

5.2. Selection Constraint CP: Constant Population

For the following discussion some abbreviations will be of use. We shall call the parameter

$$E_i = A_i - D_i \tag{5.15}$$

the *excess productivity* of the molecular species I_i and shall define the function

$$\bar{E}(t) = \frac{\sum_{i=1}^{N} E_i x_i}{\sum_{i=1}^{N} x_i} \tag{5.16}$$

as the *average excess productivity*.

We shall further denote the quantity

$$W_i^{CP} = A_i Q_i - D_i \tag{5.17}$$

as the *selection value* of the information carrier I_i under CP conditions.

We now impose upon our system the CP constraint, by keeping the total population of molecular species constant. For this purpose the global dilution flux Φ_0 will be adjusted so as to keep pace with - and compensate at all times - the increase in the total polymer concentration. Φ_0 must therefore satisfy the condition

$$\Phi_0 = \sum_{i=1}^{N} E_i x_i \tag{5.18}$$

It follows from this, in conjunction with Eq. (5.8) and the definition (5.16), that at any time t

$$\varphi \equiv \overline{E}(t) \tag{5.19}$$

The set of equations (5.10) thus changes under CP conditions into

$$\dot{x}_i = (W_i^{CP} - \overline{E})x_i + \sum_{j \neq i} \psi_{ij} x_j \qquad (i,j = 1,\ldots,N) \tag{5.20}$$

These are the fundamental phenomenological equations which describe selection among biological macromolecules under the constraint of constant population.

The CP selection equations are coupled to each other by their common quantity $\overline{E}$, which is by definition a function of all x_j. Thus the selection equations are inherently non-linear, even if all the parameters W_i^{CP} are independent of the x_j. However, in general the W_i^{CP} may not only depend upon the concentration variables x_k and m_μ, but may also depend explicitly upon time. The latter case becomes relevant if a time dependence of the surroundings (*i.e.*, time dependence of the boundary conditions such as temperature *etc.*) is to be included in our model system. To simplify discussion we shall from now on assume, unless otherwise stated, that in CP systems the parameters W_i^{CP} are constant.

Before attempting a complete analytical solution of the CP selection equations, we shall discuss a special case, which will demonstrate some of the characteristic features of molecular selection and evolution.

We shall assume that all molecular species can reproduce themselves without error, so that $Q_i = 1$ and $\psi_{ij} = 0$. Equations (5.20) then become

$$\dot{x}_i = (W_i^{CP} - \overline{E})x_i \qquad (W_i^{CP} \equiv E_i) \tag{5.21}$$

The structure of these equations reveals at once some qualitative properties of the solution. Thus under CP conditions only those species increase in concentration whose selection value lies above the threshold value $\bar{E}$:

$$W_i^{CP} > \bar{E}(t) \tag{5.22}$$

All species for which $W_i^{CP} < \bar{E}(t)$ die out. As a consequence of this segregation process, the threshold value of $\bar{E}(t)$ increases continuously until a plateau is reached. At this stage the only species left is the one with the highest W_i^{CP} value, so that

$$\lim_{t\to\infty} \bar{E}(t) = W_m^{CP} \qquad \text{with} \qquad W_m^{CP} = \max_i W_i^{CP} \tag{5.23}$$

In the limit the rate of change of x_m is zero ($\dot{x}_m = 0$). The system is now in a stationary state, which we shall call *selection equilibrium*. A computer simulation of such a selection competition is shown in Fig. 5.1. It confirms the conclusions to which our qualitative discussion has lead us.

Another means of confirmation is provided by a *linear stability analysis* of the stationary states, which is explained in Appendix A.2 and will be used later for more complex cases. In this instance N different stationary states are conceivable, in each of which one of the N species has displaced all the others:

$$\mathbf{x}_1^s = \begin{pmatrix} c \\ 0 \\ \vdots \\ \vdots \\ \vdots \\ 0 \end{pmatrix}, \dots, \mathbf{x}_k^s = \begin{pmatrix} 0 \\ \vdots \\ 0 \\ c \\ 0 \\ \vdots \\ 0 \end{pmatrix}, \dots, \mathbf{x}_N^s = \begin{pmatrix} 0 \\ \vdots \\ \vdots \\ \vdots \\ 0 \\ c \end{pmatrix} \tag{5.24}$$

Each stationary state S_k can thus be described by an N-tuplet of concentration values. For the stationary state S_k the relation

$$\bar{E} = W_k^{CP} \tag{5.25}$$

applies. The linear stability analysis yields the following spectrum of eigenvalues:

$$\lambda_k = -W_k^{CP} \quad \text{and} \quad \lambda_j = W_j^{CP} - W_k^{CP} \quad \text{for all} \quad j \neq k \tag{5.26}$$

However, a stationary state is stable if and only if all its eigenvalues have negative real parts. Thus the distribution of eigenvalues (5.26) gives only one stable stationary state, that is, the state S_m, which corresponds to the greatest selection value W_m^{CP}.

A molecular selection process with kinetics described by Eq. (5.21) thus always leads to an unambiguous selection "decision" - to the survival of the best-fitted species and to the extinction of all its competitors. It is a consequence of the particular structure of Eq. (5.21) that the asymptotically stable solution at the same time represents the sole stable stationary state. However, other cases are also possible in which the concentration variables oscillate, so that asymptotically stable, closed paths in phase space (periodic solutions) are found. Such behaviour occurs only in systems in which certain kinds of feedback between the species are present. In Part III we shall make the acquaintance of some cases of this kind.

We must not forget that Eq. (5.21) is a special case of Eq. (5.20), in which the possibility of mutation has been omitted. In real systems, however, Q_i is necessarily less than unity (for all species I_i), so that the species I_m which is selected is always surrounded by a stationary distribution of mutants. The complete stationary distribution is termed a *quasi-species*, with the *master sequence* I_m. A newly arising mutant I_{m+1} with a selection value $W_{m+1}^{CP} > W_m^{CP}$ will (within the limits of the deterministic approach) displace its master sequence I_m and will itself dominate in the selection equilibrium. This process can be equated with the evolutionary progress of the system.

The inherent tendency of the copying process to produce mutants will in general upset the attainment of selection equilibrium. If however the mutation rates are

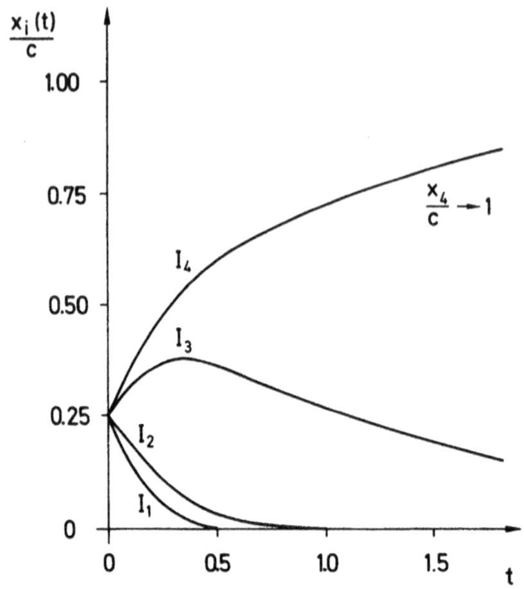

Fig. 5.1. Competition between four species in a CP system (without mutation). All species start at time t=0 with the same concentration: $x_1(0) = x_2(0) = x_3(0) = x_4(0) = c/4$. Selection takes place in accordance with Eq. (5.21) and the parameter values $W_1^{CP} = 1$, $W_2^{CP} = 4$, $W_3^{CP} = 9$, and $W_4^{CP} = 10$.

At time t=0 the average excess productivity has a value of $\bar{E} = 6$. Therefore, from the start, the concentrations of species I_1 and I_2 will decrease. I_3 and I_4 grow to begin with. Later the average excess productivity exceeds the selection value of species I_3 and this species becomes extinct along with I_2 and I_1. (After Eigen, 1971.)

small enough ($Q_i \lesssim 1$ and $\psi_{ij} \ll 1$) then the time scale of selection becomes much shorter than that of evolution.

A characteristic measure of the rate of attainment of an equilibrium is the so-called *relaxation time*. This is the time constant with which a perturbation of a stable equilibrium decays exponentially (*cf.* Appendix A.2).

The relaxation time $\tau_{\text{selection}}$ is given roughly by the reciprocal of the smallest eigenvalue from Eq. (5.26), *i.e.*,

$$\tau_{\text{selection}} \approx \frac{1}{W_m^{CP} - W_{m-1}^{CP}} \quad (W_m^{CP} > W_{m-1}^{CP}) \tag{5.27}$$

where W_m^{CP} and W_{m-1}^{CP} are the largest selection values among the initial distribution.

This is also seen in Fig. 5.1; after $t = (W_4^{CP} - W_3^{CP})^{-1} = 1$ the result of the selection competition has become clear.

In selection equilibrium the dominant species I_m is finally competing only with its own incorrect copies $I_{k \neq m}$. These arise from the master sequence with a production rate of

$$\frac{\Delta x_k}{\Delta t} \approx \psi_{km} x_m \approx \psi_{km} c \tag{5.28}$$

so that the order of magnitude of waiting time before a selectively more favourable mutant arises is given by

$$\tau_{\text{mutation}} = \frac{1}{\psi_{m+1,m}\, Z} \tag{5.29}$$

where I_{m+1} is the superior mutant most likely to arise from the master sequence I_m and Z is the population number.

As long as $\tau_{\text{selection}} \ll \tau_{\text{mutation}}$, Eq. (5.21) provides a good description of the selection process. The process of evolution is then given by the rate equation for the formation of better-equipped mutants I_{m+v} from the master sequence I_m:

$$\dot{x}_{m+v} = (W_{m+v}^{CP} - \overline{E})x_{m+v} + \psi_{m+v,m}\, x_m \qquad (v = 1,2...) \tag{5.30}$$

Here we have made use of the fact that in selection equilibrium $\overline{E} \approx W_m^{CP}$.

The stationary solutions of Eqs. (5.20) and (5.30) can easily be given for $Q_m \lesssim 1$ and $\psi_{ij} \ll 1$. For this purpose let us divide $\overline{E}$ into the two parts which arise respectively from the master sequence I_m and from the mutant pool $I_{k \neq m}$:

$$\overline{E} = \overline{E}_{k \neq m} + \frac{x_m}{c}(E_m - \overline{E}_{k \neq m}) \tag{5.31}$$

where the average productivity of the distribution of mutants is given by

$$\bar{E}_{k\neq m} = \frac{\sum_{k\neq m} E_k x_k}{\sum_{k\neq m} x_k} \tag{5.32}$$

If we insert the expression (5.32) for $\bar{E}$ into Eq. (5.21) we obtain

$$\dot{x}_m = (W_m^{CP} - \bar{E}_{k\neq m})x_m - \frac{x_m^2}{c}(E_m - \bar{E}_{k\neq m}) \tag{5.33}$$

from which, with the condition $\dot{x}_m = 0$, the stationary-state solution

$$x_m^s = c\,\frac{W_m^{CP} - \bar{E}_{k\neq m}}{E_m - \bar{E}_{k\neq m}} \approx c \qquad (W_m^{CP} \approx E_m) \tag{5.34}$$

follows. Similarly Eq. (5.30) leads with the condition $\dot{x}_{m+v} = 0$ to the stationary-state solution for the mutant I_{m+v}:

$$\left(\frac{x_{m+v}}{x_m}\right)^s = \frac{\psi_{m+v,m}}{W_m^{CP} - W_{m+v}^{CP}} \approx 0 \qquad (\psi_{m+v,m} << W_m^{CP} - W_{m+v}^{CP}) \tag{5.35}$$

whereby we have made use of the fact that in selection equilibrium $\bar{E} \approx E_m \approx W_m^{CP}$.

The above considerations lead to the conclusion that the overall process of molecular self-organization with low rates of mutation takes place on two time-scales, one fast and one slow:

(1) SELECTION (fast). A particular initial distribution of molecular species is narrowed down to one species, the one with the highest selection value W_m^{CP}. The "selection equilibrium" consists of the dominant species I_m (master sequence) and a stationary distribution of mutants arising from it by a fallible copying process. This "equilibrium" includes possible fluctuations in the error distribution and is metastable with respect to the appearance of mutants with a selective advantage.

(2) EVOLUTION (slow). If a mutant I_{m+v} with a selective advantage ($W_{m+v}^{CP} > W_m^{CP}$) appears, then the equilibrium breaks down and within the relaxation time $\tau_{selection}$

a new equilibrium is reached, this time dominated by species I_{m+v}. The transition from one selection equilibrium to another also optimizes the equilibrium value of $\bar{E}$, in the sense that $\bar{E}$ reaches the absolute maximum which the system can attain. For quasi-linear systems as in Eq. (5.21), this optimization process is a strictly monotonic growth of W_m^{CP}: the Darwinian concept of "fitness" thus appears as a parameter whose value follows directly from material properties of the system (such as expressed by Definition 5.17). The "fittest" species can be regarded as the species with the highest value of this parameter. In accordance with the inequality

$$\ldots\ W_{m-1}^{CP} < W_m^{CP} < W_{m+1}^{CP} < \ldots < W_{opt}^{CP} \tag{5.36}$$

the progress of the system is confined to a particular path, along which it organizes itself spontaneously. Consequently, of the N possible molecular states, only a restricted number of states is passed through.

The relation (5.36) is, in the simple form given, true only of *quasi-linear* systems, *i.e.*, those in which the rate constants are independent of the concentration variables. In the general case of a non-linear system the optimization by no means reflects a simple, monotonic change in selection values. As soon as internal feedback arises among the species, the selection values become concentration-dependent and the entire optimization process becomes more complicated. In particular, the term "highest selection value" is in this case meaningless, since it now depends on the other molecular species present. Every change in the concentration variables x_k becomes a change in the environment of other species and thus a change in their selection values. The optimization is in this case a "differential" process. Thus, for example, the selection criterion (5.23) for oscillating systems is

$$\lim_{t\to\infty} \langle \bar{E} \rangle = \langle W_m^{CP} \rangle \tag{5.37}$$

where $\langle \bar{E} \rangle$ and $\langle W_m^{CP} \rangle$ are the average productivity and the selection value averaged over one oscillation cycle Δt, *i.e.*,

$$\langle \bar{E} \rangle = \frac{1}{\Delta t} \int_t^{t+\Delta t} \bar{E} \, d\tau \quad \text{and} \quad \langle W_m^{CP} \rangle = \frac{1}{\Delta t} \int_t^{t+\Delta t} W_m^{CP} \, d\tau \tag{5.38}$$

5.3. Solution of the CP Selection Equations

Even though we were able to deduce some of the essential features of selection from the approximate form (5.21) of the CP selection equations, we shall now need an exact solution of these equations in order to see in more depth the principles involved. To do this we first cast the Eqs. (5.20) into a somewhat different form:

$$\dot{x}_i = \sum_j w_{ij} x_j - \bar{E}\, x_i \qquad (i,j = 1,\ldots,N) \tag{5.39}$$

whereby the coefficients w_{ij} are defined as the elements of the matrix

$$\mathbf{W} = (w_{ij}) = \begin{cases} W_i^{CP} & \text{for } i=j \\ \psi_{ij} & \text{for } i \neq j \end{cases} \tag{5.40}$$

We have already mentioned that these equations are inherently non-linear, even when the coefficients w_{ij} are constant. Because of this non-linearity, an analytical solution is only possible if all the coefficients w_{ij} are independent of the concentration variables x_k (which however in general may not be the case). We shall therefore assume that the w_{ij} *are* independent of the x_k and shall use this assumption to turn Eqs. (5.39) into a set of linear equations. These are easily solved if we further assume that the w_{ij} are also independent of time.

In order to solve Eqs. (5.39) we first introduce a new variable $q_i = q_i(t)$, defined by

$$x_i(t) = h(t)q_i(t) \qquad (i = 1,\dots,N) \tag{5.41}$$

where

$$h(t) = \exp\left(-\int_0^t \bar{E}(\tau)\, d\tau\right) \tag{5.42}$$

If this transformation is substituted into Eqs. (5.39), the result is a set of linear differential equations:

$$\dot{q}_i = \sum_j w_{ij}\, q_j \qquad (i,j = 1,\dots,N) \tag{5.43}$$

or, written in vector notation,

$$\dot{\mathbf{q}} = \mathbf{W}\mathbf{q} \tag{5.44}$$

With the help of the transformation (5.41/5.42) we have thus reduced the solution of Eqs. (5.39) to a simple eigenvalue problem (*cf.* Appendix A.1). The corresponding characteristic equation runs

$$\det(\mathbf{W} - \lambda \mathbf{I}) = 0 \tag{5.45}$$

where $\mathbf{I}$ is the *identity matrix* and λ are the eigenvalues of the matrix $\mathbf{W}$. The solution of the characteristic equation yields a spectrum of N eigenvalues λ_i, each associated with a normal mode v_i of the reaction system. These normal modes are linear combinations of the concentration variables of all the species involved in the overall reaction. They satisfy (by definition) the relation

$$\dot{\mathbf{v}} = \mathbf{L}\mathbf{v} \tag{5.46}$$

where $\mathbf{L}$ is a diagonal matrix with values λ_i on the diagonal, *i.e.*,

$$\mathbf{L} = (l_{ij}) = \begin{cases} \lambda_i & \text{for } i=j \\ 0 & \text{for } i \neq j \end{cases} \tag{5.47}$$

The relation between the $\mathbf{v}$- and the $\mathbf{q}$-variables is given by

$$\mathbf{q} = \mathbf{U}\mathbf{v} \qquad \text{or} \qquad \mathbf{v} = \mathbf{U}^{-1}\mathbf{q} \tag{5.48}$$

where the transformation matrix $\mathbf{U} = (u_{ij})$ and its inverse $\mathbf{U}^{-1}$ are determined by the components of the eigenvectors of $\mathbf{W}$. Since $\mathbf{L}$ is a diagonal matrix Eq. (5.46) can at once be solved, and, by use of the linear transformation (5.48), a solution for $\mathbf{q}$ and thus for $\mathbf{x}$ can be found.

The formal solution of Eqs. (5.20) is thus complete. Some important physical aspects of the CP selection equations will now emerge as we carry out the solution in practice.

By adding up all the equations (5.39) we obtain

$$\sum_i \dot{x}_i = \sum_i \sum_j w_{ij} x_j - \bar{E} \sum_i x_i \tag{5.49}$$

Bearing in mind the CP constraint (5.13) we can immediately write down the following formula for $\bar{E}$:

$$\bar{E} = \frac{\sum_i \sum_j w_{ij} x_j}{\sum_i x_i} \tag{5.50}$$

Into this we substitute the transformation (5.41) and obtain

$$\bar{E}(t) = \frac{\sum_i \sum_j w_{ij} q_j}{\sum_i q_i} \tag{5.51}$$

Adding up Eqs. (5.43) we find

$$\sum_i \dot{q}_i = \sum_i \sum_j w_{ij} q_i \tag{5.52}$$

and this becomes, applying Eq. (5.51)

$$\sum_i \dot{q}_i = \bar{E}(t) \sum_i q_i \tag{5.53}$$

which is equivalent to

$$\bar{E}(t) = \frac{\partial}{\partial t} \ln \sum_i q_i(t) \tag{5.54}$$

Integrating, we obtain

$$\sum_i q_i(t) = \alpha \exp \left(\int_0^t \bar{E}(\tau)\, d\tau \right) \tag{5.55}$$

where α is a constant of integration. Comparison with the transformation (5.41/5.42) shows that

$$h(t) = \frac{\alpha}{\sum_i q_i(t)} \tag{5.56}$$

The constant of integration is determined by the initial conditions. At time t=0 we have h(0) = 1, so that

$$\alpha = c \tag{5.57}$$

Substituting Eq. (5.56) into Eqs. (5.41) we can thus express the solution of Eqs. (5.39) as

$$x_i(t) = c \frac{q_i(t)}{\sum_j q_j(t)} \qquad (i,j = 1,\ldots,N) \tag{5.58}$$

where $q_i(t)$ is determined by Eqs. (5.43).

Equations (5.43), which will lead to the functions $q_i=q_i(t)$, can as stated above be solved with the aid of simple algebraic procedures (see Appendix A.1). For this purpose we start from the eigenvalue equation

$$\sum_j w_{ij}u_{jk} = \lambda_k u_{ik} \tag{5.59}$$

In this equation λ_k are the eigenvalues of the matrix $\mathbf{W}$ and u_{ik} are the components of the corresponding eigenvectors. The eigenvalues λ_k are solutions of the characteristic equation

$$\det\,(w_{ij}-\lambda\delta_{ij}) = 0 \tag{5.60}$$

where δ_{ij} is the *Kronecker Delta*, defined by

$$\delta_{ij} = \begin{cases} 1 & i=j \\ 0 & i\neq j \end{cases} \tag{5.61}$$

A general solution of Eq. (5.43) can be given immediately if all of the eigenvalues defined by Eq. (5.59) are *non-degenerate*, *i.e.*, no two of them are equal. In this case all the eigenvalues are determined unambiguously, and the solution of Eq. (5.43) is

$$q_i(t) = \sum_k u_{ik}\, a_k\, \exp(\lambda_k t) \tag{5.62}$$

Thus we obtain as the complete, time-dependent solution of the CP selection equations (5.20) or (5.39) for non-degenerate eigenvalues λ_k

$$x_i(t) = c\,\frac{\sum_k u_{ik}\, a_k\, \exp(\lambda_k t)}{\sum_j \sum_k u_{jk}\, a_k\, \exp(\lambda_k t)} \qquad (i,j = 1,\ldots,N) \tag{5.63}$$

The constants a_k can be found from the initial conditions. From Eq. (5.63), at t=0, follows

$$a_k = \sum_l u_{kl}^{-1} x_l(0) \tag{5.64}$$

where u_{kl}^{-1} are the elements of the inverse matrix $\mathbf{U}^{-1}$.

One can make certain general statements about the properties of the solutions (5.63). If the mutation matrix (ψ_{ij}) is symmetric (that is, $\psi_{ij} = \psi_{ji}$ for all i and j) then the matrix $\mathbf{W}$ is *Hermitian* and all the eigenvalues λ_k are real. In general, however, the symmetry condition is not met, and the eigenvalues may be complex. This would in turn mean that the concentration variables x_i oscillate in time.

For real biological systems the mutation parameters ψ_{ij} are always greater than zero, since for basic physical reasons no nucleic acid can reproduce itself perfectly ($Q_i < 1$) and no possible kind of error is excluded, so that formation of any I_i by mutation of any I_j has a finite probability. In this case complex eigenvalues only appear when the off-diagonal elements of $\mathbf{W}$ are relatively large and of the same order of magnitude as the diagonal elements.

Equation (5.63) represents a general solution of the CP selection equations. In principle, the eigenvalues λ_j and the components u_{ij} of the corresponding eigenvectors can be calculated explicitly for any given matrix $\mathbf{W}$ with constant coefficients by the methods described in the Appendix A.1. However, for large numbers of competing species the eigenvalue problem becomes hopelessly complicated. In such cases we have to use suitable approximative methods to calculate the eigenvalues and eigenvectors.

For biological information carriers ψ_{ij} is as a rule small in comparison with W_i^{CP}, so that we can evaluate the eigenvalues and eigenvectors with the help of second-order perturbation theory. This yields for the eigenvalues λ_i and the components u_{ij} of the eigenvectors the following approximate expressions:

$$\lambda_j \approx W_j^{CP} + \sum_{i \neq j} \frac{\psi_{ij}\,\psi_{ji}}{W_j^{CP} - W_i^{CP}} \tag{5.65}$$

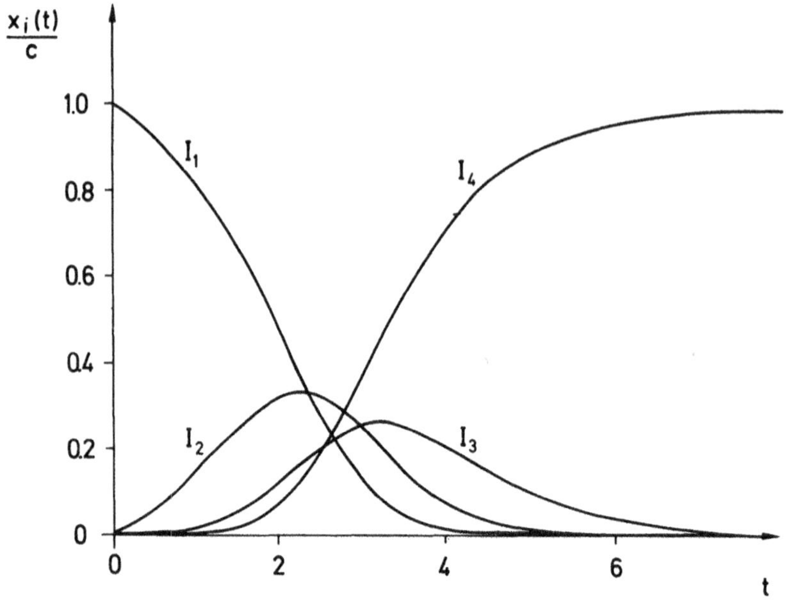

Fig. 5.2. Competition between four species in a CP system (small mutation terms). At time t=0 species I_1 dominates completely: $x_1(0)=c$, $x_2(0)=x_3(0)=x_4(0)=0$. The concentration distribution changes in accordance with Eq. (5.39), where

$$\mathbf{W} = \begin{pmatrix} 1 & 0.001 & 0.001 & 0.01 \\ 0.1 & 2 & 0.01 & 0.01 \\ 0.001 & 0.1 & 3 & 0.01 \\ 0.001 & 0.001 & 0.001 & 4 \end{pmatrix}$$

All mutation terms ψ_{ij} are small in comparison with the selection values W_i^{CP}. (After Jones *et al.*, 1976a.)

and

$$u_{ij} \approx \delta_{ij} + \frac{\psi_{ij}}{W_j^{CP}-W_i^{CP}} \tag{5.66}$$

For constant coefficients w_{ij} the eigenvalues λ_j thus depend upon the inherent selection values W_i^{CP} of all the species present in the system. As we shall show in more detail in Sections 5.4 and 5.6, the parameters λ_j represent generalized

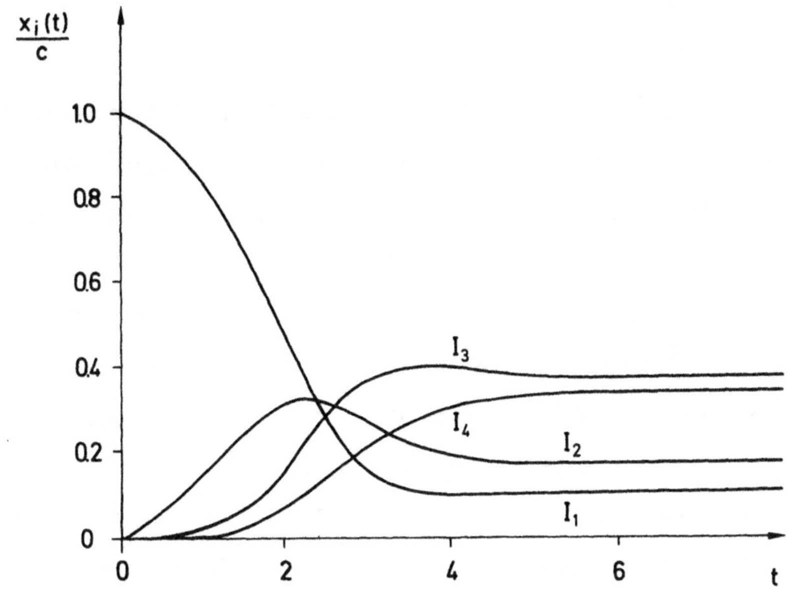

Fig. 5.3. Competition between four species in a CP system (large mutation terms). At time t=0 species I_1 dominates completely over the system: $x_1(0)=c$, $x_2(0)=x_3(0)=x_4(0)=0$. The distribution changes in accordance with Eq. (5.39), where

$$\mathbf{W} = \begin{pmatrix} 1 & 0.001 & 0.001 & 1 \\ 0.1 & 2 & 0.01 & 1 \\ 0.001 & 0.1 & 3 & 1 \\ 0.001 & 0.001 & 0.001 & 4 \end{pmatrix}$$

The mutation terms ψ_{14} , ψ_{24} and ψ_{34} are of the same order of magnitude as the selection values W_i^{CP}. (After Jones *et al.*, 1976a.)

selection values which are characteristic for the selection behaviour of the system as a whole.

The discussion of Eq. (5.21) and Fig. 5.1 has made it clear that without mutation terms ψ_{ij} there is always an unambiguous selection, that is, survival of only one species. We shall now consider two cases in which as a consequence of finite mutation terms ψ_{ij} the initial distribution breaks down and is replaced by a new distribution dominated by newly-arisen mutants with a selective advantage. In both cases the initial conditions for competition are the same: $x_1(0)=c$ and $x_2(0)=x_3(0)=x_4(0)=0$.

In the first case (Fig. 5.2) it is assumed that the mutation terms ψ_{ij} are all small in comparison with the W_i^{CP} terms. The computer simulation shows that here again the result is an unambiguous selection of the species with the highest selection value (I_4).

In the second case (Fig. 5.3) some of the ψ_{ij} terms are of the same order of magnitude as the W_i^{CP} terms. The species I_4 still has the greatest selection value, but it cannot assert itself unambiguously on account of its high mutation rate (W_4^{CP} and the ψ_{i4} are similar in magnitude). Because of the relatively high selection value W_3^{CP}, the large mutation term ψ_{34} of I_4 to I_3 and the small mutation terms ψ_{i3}, the stationary-state concentration of the species I_3 is greater than that of species I_4.

In cases where the coefficients w_{ij} are functions of time, a complete and exact solution of Eq. (5.43) is not under all circumstances possible. Here again we must recourse to an approximate solution. Thus Jones (1977) was able to derive exact asymptotic solutions for time-dependent coefficients w_{ij} with a particular form in the limit $t \rightarrow \infty$.

5.4. The Quasi-Species

The mathematical procedure described in the preceding section for the solution of the CP selection equations consists in essence of two steps. In the first step the non-linear set of differential equations (5.39) was transformed into a linear one (Eq. 5.43). In the second step the transformed set of equations (5.43) was treated as a simple eigenvalue problem, that is, the matrix $\mathbf{W}$ was converted to a diagonal form by an affine transformation.

It is also possible to invert the order of the two steps and begin with the second. Although these procedures are completely equivalent, the second will lead us to an important concept in selection.

Let us start with Eqs. (5.10) and assume to begin with that there is *no* population control in our model system ($\varphi \equiv 0$). We shall also need to "re-classify" the polymers. We define a set of N *quasi-species* whose concentration variables v_j are linear combinations of the original variables x_i, that is,

$$x_i = \sum_j u_{ij} v_j \qquad \text{or} \qquad v_i = \sum_j u_{ij}^{-1} x_j \tag{5.67}$$

We shall further choose the coefficients u_{ij} such that after use of the transformation (5.67) Eqs. (5.10) take on the form:

$$\dot{v}_j = \lambda_i v_i \qquad (i = 1,\ldots,N) \tag{5.68}$$

In other words, we are again dealing with an affine transformation. The λ_i are again the eigenvalues of the matrix **W** and the coefficients u_{ij} again represent the ith components of the eigenvector corresponding to the eigenvalue λ_j.

We now introduce a population control adjusted so as to give exactly the CP constraint. Applying the transformation (5.67) to the differential equations (5.20) we obtain

$$\dot{v}_i = (\lambda_i - \bar{E}(x_1,\ldots,x_N))v_i \qquad (i = 1,\ldots,N) \tag{5.69}$$

Application of the CP constraint means that

$$\sum_i x_i = \sum_i \sum_j u_{ij} v_j = \sum_j (\sum_i u_{ij})\, v_j = c = \text{const.} \tag{5.70}$$

We now normalize the components u_{ij} of the jth eigenvector with the constants $\sum_i u_{ij}$. In doing this we take account of the physical fact that re-grouping the polymers into quasi-species leaves the total concentration of each molecular species unaltered. Thus the CP condition still applies, *viz.*:

$$\sum_i x_i = \sum_j v_j = c \tag{5.71}$$

We now add up all the Eqs. (5.69) and obtain, applying the condition (5.71),

$$\dot{c} = (\bar{\lambda}-\bar{E})c \qquad \text{where} \qquad \bar{\lambda} = \frac{1}{c}\sum_j \lambda_j v_j \tag{5.72}$$

In the CP system, however, we have $\dot{c}=0$, so that for non-trivial solutions of Eqs. (5.72) (*i.e.*, those with $c \neq 0$) the identity

$$\bar{\lambda} \equiv \bar{E} \tag{5.73}$$

must hold.

The quantity $\bar{E}$ is thus unaffected by the transformation and can be expressed as the average of all the eigenvalues λ_j. From Eqs. (5.69) it follows that the selective self-reproduction of a population of N quasi-species is then described by the following set of differential equations:

$$\dot{v}_i = (\lambda_i-\bar{\lambda})v_i \qquad (i = 1,\dots,N) \tag{5.74}$$

When errors begin to occur in the reproduction process, the eigenvalues λ_i become dependent upon the inherent selection values W_j^{CP} of all the other species I_j. The eigenvalues λ_i therefore represent generalized value parameters, which characterize the collective selection behaviour of the molecular species. We have already seen this result in discussing the approximate solution (5.65).

Notice also the structural similarity between Eqs. (5.74) and (5.21). By analogy with our discussion of Eq. (5.21) we can anticipate the characteristics of selection implicit in Eqs. (5.74). All the selection equations are coupled to each other by the average eigenvalue $\bar{\lambda}$. This represents a sliding threshold value, which reflects the degree of self-organization of the system. Only those quasi-species increase in concentration whose eigenvalues λ_i lie above the threshold value $\bar{\lambda}$. In consequence of their growth, the threshold value is steadily shifted to higher values until a maximum is reached, equal to the highest eigenvalue of the distribution:

$$\lim_{t\to\infty} \bar{\lambda}(t) = \lambda_m \qquad (\lambda_m = \max_i \lambda_i) \tag{5.75}$$

In the limit (Eq. 5.75) all the quasi-species except for the one with the highest eigenvalue have died out.

Let us examine more closely the physical meaning of the concept of a "quasi-species". In biology the meaning of "species" is well-known: it embraces a class of individuals which have certain *phenotypic* properties in common. None the less, we know that the individual members of a species are usually *genotypically* different. The differences in the DNA or RNA within a species are as a rule very slight, so that the sequences involved are scattered narrowly around a defined average sequence. The optimally adapted phenotype which possesses this sequence within a distribution of mutants is in biology called the *wild-type* of the population. However, sequence analysis of phage populations has shown that in practice only a small fraction of the nucleic acid sequences is identical to the wild-type sequence (Batschelet *et al.*, 1976). Most of the individuals differ from the wild-type in one or more point mutations. Thus even in a state of selection equilibrium, a population is - because of the finite error rate in the copying of genetic information - never homogeneous, but instead is always a distribution of mutants which have arisen from a common wild-type. Selection thus never operates upon a uniform class of individuals, but rather upon a distribution of this kind. Thus the mathematical concept of a "quasi-species" fits in perfectly with the realities of molecular biology.

Solving the differential equations (5.74) is no problem. We start with the implicit solution which is obtained by direct integration of Eq. (5.74):

$$v_i(t) = \frac{v_i(0)\ \exp(\lambda_i t)}{\exp(\int_0^t \bar{\lambda}(\tau)d\tau)} \tag{5.76}$$

The integral term cancels out in the ratio v_i/v_j, which we then insert into the definition of $\bar{\lambda}$ (see Eq. 5.72):

$$\bar{\lambda} = \frac{v_i}{c} \sum_j \frac{v_j}{v_i} \lambda_j \tag{5.77}$$

Equations (5.74) may then be written as

$$\dot{v}_i = \lambda_i v_i - \frac{v_i^2}{c v_i(0)} \sum_j v_j(0) \lambda_j \exp(\lambda_j - \lambda_i)t \tag{5.78}$$

This is a form of the *Bernoulli equation*. If we define

$$g(t) = \lambda_i \tag{5.79}$$

and

$$h(t) = \frac{1}{c v_i(0)} \sum_j v_j(0) \lambda_j \exp(\lambda_j - \lambda_i)t \tag{5.80}$$

then Eq. (5.78) has the general form

$$\frac{1}{v(t)} = e(t) \int \frac{h(\tau)}{g(\tau)} d\tau \qquad \text{with} \qquad e(t) = \exp\left(\int g(\tau) d\tau\right) \tag{5.81}$$

It is finally a simple matter to obtain the explicit solutions of Eqs. (5.74), which are

$$v_i(t) = \frac{v_i(0) c \exp(\lambda_i t)}{\sum_j v_j(0) \exp(\lambda_j t)} \qquad (i,j = 1,\ldots,N) \tag{5.82}$$

We now insert the solutions (5.82) for the normal modes $v_i = v_i(t)$ into Eq. (5.67) and obtain as solutions for the concentration variables x_i

$$x_i(t) = \frac{c \sum_j u_{ij} \alpha_j \exp(\lambda_j t)}{\sum_j \sum_i u_{ij} \alpha_j \exp(\lambda_j t)} \qquad (i,j = 1,\ldots,N) \tag{5.83}$$

The coefficients a_j are obtained by rearranging Eqs. (5.83) at t=0.

As expected, the solutions (5.83) are identical with the solutions (5.63), although we must bear in mind that in the second case we have normalized each eigenvector $\mathbf{u}_j$ with the quantity $\sum_i u_{ij}$.

At $t \rightarrow \infty$ the selection equilibrium is reached and we obtain from Eq. (5.83) for the dominant species I_m

$$\lim \frac{x_i}{c} = \frac{u_{im}}{\sum_i u_{im}} \tag{5.84}$$

Thus in selection equilibrium the proportion of each species I_i is equal to its proportion in the selected quasi-species. The same result will be obtained for the selection kinetics in CF systems (Eq. 5.183).

5.5. Selection Dynamics in CP Systems

The foregoing discussion has shown that the overall process of molecular self-organization is described basically by the threshold function $\bar{\lambda}$. If particular prerequisites concerning the coefficients w_{ij} are fulfilled, then further conclusions about the time-dependence of $\bar{\lambda}$ can be reached (Küppers, 1979a). In the following discussion we shall assume that the elements of the matrix $\mathbf{W}$ are constant and are not negative, and in addition that all the normal modes v_i are positive at time t=0. It follows from these conditions that for $t \geq 0$ all $v_i(t)$ are real and positive (Bellman, 1970). The physical meaning of these assumptions has already been discussed at length in Section 5.3.

We consider the change in time of $\bar{\lambda}$ (*cf.* Definition 5.72)

$$\dot{\bar{\lambda}} = \frac{1}{c} \sum_i \lambda_i \dot{v}_i \tag{5.85}$$

Applying Eqs. (5.74) we obtain first of all

$$\dot{\bar{\lambda}} = \frac{1}{c} \sum_i \lambda_i(\lambda_i-\bar{\lambda})v_i = \sum_i (\lambda_i-\bar{\lambda})^2 \frac{v_i}{c} \tag{5.86}$$

The right-hand expression is recognizable as the variance V of λ_i. Since all the squared terms and concentrations are positive, it follows that

$$V = \dot{\bar{\lambda}} = \overline{\lambda^2} - \bar{\lambda}^2 \geq 0 \tag{5.87}$$

That is to say, the rate of change in time of $\bar{\lambda}$ is at every instant equal to the variance V of the eigenvalues λ_i.

In accordance with relation (5.87), the effectiveness of displacement of the distribution of concentrations by selection depends upon the variance of the reproduction parameters of all the species taking part in the selection process. Away from selection equilibrium we have $\dot{\bar{\lambda}} > 0$ and at selection equilibrium we have $\dot{\bar{\lambda}} = 0$. Equation (5.87) is formally equivalent to R.A. Fisher's fundamental theorem of natural selection (Fisher, 1930) and thus provides a link between the selection models decribed here and those of population biology (*cf.* in particular Crow and Kimura, 1970; Davies, 1978).

Next let us consider the time-dependence of the variance V. Starting from Eq. (5.86) we obtain

$$\dot{V} = \sum_i (\lambda_i-\bar{\lambda})^2 \frac{\dot{v}_i}{c} - 2 \sum_i (\lambda_i-\bar{\lambda})\dot{\bar{\lambda}} \frac{v_i}{c} = \sum_i (\lambda_i-\bar{\lambda})^3 \frac{v_i}{c} \tag{5.88}$$

The rate of change of the variance V is thus equal to the third moment of the eigenvalue distribution. The third moment is a fundamental measure of the asymmetry of a distribution. A fully symmetric distribution of the eigenvalues λ_i about their mean $\bar{\lambda}$ implies that $\dot{V}$ is zero and that $\bar{\lambda}$ changes at a constant rate. This case is however unrealistic, since in a random (*e.g.* prebiotic) distribution of nucleic acids, the eigenvalues at the beginning of a selection process are as a rule *not* symmetrically scattered around the mean, $\bar{\lambda}$.

We are now in a position to prove an important theorem concerning the rate of molecular self-organization under the CP constraint:

THEOREM: *With a given variance in a short time interval the* v*-variables change in such a way that the rate of increase in the mean eigenvalue of the population is maximized.*

In other words, the selection equilibrium is reached along a path for which the line integral

$$I = \int_1^2 d\bar{\lambda} \qquad (5.89)$$

(for fixed limits t_1 and t_2) is zero, *i.e.*,

$$\delta I = \delta \int_{t_1}^{t_2} \dot{\bar{\lambda}} \, dt = 0 \qquad (5.90)$$

To illustrate Eq. (5.90) we shall show that the maximization of $\dot{\bar{\lambda}}$ leads to the selection equations (5.74). This procedure is similar to the well-known procedure in clasical mechanics by means of which the Newtonian equations of motion can be derived from the Hamiltonian integral principle.

In order to maximize $\dot{\bar{\lambda}}$ we make use of the method of Lagrange multipliers (*cf.* Appendix A.3). This is a procedure for the determination of the extreme value of a function under given constraints. In our case the first constraint comes from the CP condition (5.71):

$$\sum_i \delta v_i = 0 \qquad (5.91)$$

The second constraint emerges from the fact that the variance is (momentarily) constant. Combining Eqs. (5.74), (5.86) and (5.87) gives

$$\frac{1}{c}\sum_i \frac{1}{v_i}\left(\frac{\delta v_i}{\delta t}\right)^2 = \overline{\lambda^2} - \overline{\lambda}^2 \tag{5.92}$$

For the small increment $\delta\overline{\lambda}$, Eq. (5.72) yields

$$\delta\overline{\lambda} = \frac{1}{c}\sum_i \lambda_i \delta v_i \tag{5.93}$$

We shall now apply the usual procedure of variational calculus. We define a function G as

$$G = \delta\overline{\lambda} + \xi_1 \sum_i \delta v_i + \xi_2 \left[\overline{\lambda^2}(\delta t)^2 - \lambda^2(\delta t)^2 - \frac{1}{c}\sum_i \frac{1}{v_i}(\delta v_i)^2 \right] \tag{5.94}$$

where ξ_1 and ξ_2 are factors (Lagrange multipliers) whose value is for the present undetermined. The condition $\partial G/\partial(\delta v_i) = 0$ then yields N equations of the form

$$\frac{\lambda_i}{c} + \xi_1 + \frac{2\xi_2\, \delta v_i}{c\, v_i} = 0 \qquad (i = 1,\ldots,N) \tag{5.95}$$

We multiply the above equation by v_i and then add all the N equations together, obtaining

$$\frac{1}{c}\sum_i \lambda_i v_i + \xi_1 \sum_i v_i + \frac{2\xi_2}{c}\sum_i \delta v_i = 0 \tag{5.96}$$

Using Eq. (5.91), we can simplify Eq. (5.96) to

$$\overline{\lambda} + \xi_1 c = 0 \tag{5.97}$$

which enables us to determine the first multiplier:

$$\xi_1 = -\frac{\overline{\lambda}}{c} \tag{5.98}$$

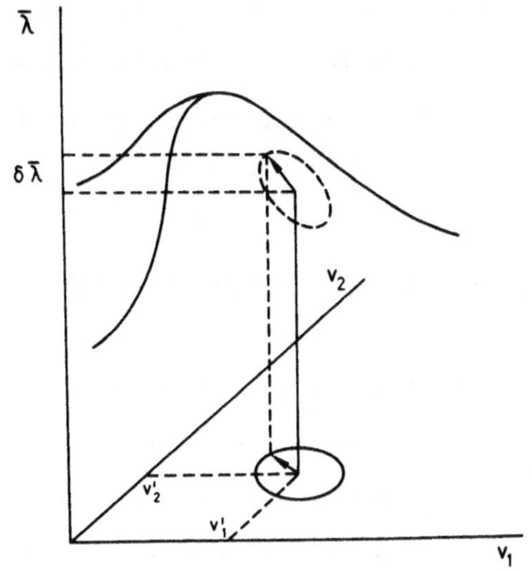

Fig. 5.4. Two-dimensional representation of the $\bar{\lambda}$-surface. In order to be compatible with the constraints of the system (Eqs. 5.91 and 5.92), each virtual displacement $(\delta v_1, \delta v_2)$ must connect the centre of an ellipse, with semiaxes $a_1 = \sqrt{v_1 c V}\,\delta t$ and $a_2 = \sqrt{v_2 c V}\,\delta t$, to its periphery.

Inserting Eq. (5.98) into Eq. (5.96) and applying Eq. (5.92) we can now find the second multiplier

$$\xi_2 = \pm \frac{1}{2\delta t} \tag{5.99}$$

Insertion of ξ_1 and ξ_2 into Eq. (5.95) and ignoring the (physically meaningless) negative root then yields as desired the familiar selection equations

$$\frac{\delta v_i}{\delta t} = (\lambda_i - \bar{\lambda}) v_i \qquad (i = 1, \ldots, N) \tag{5.100}$$

Figure 5.4 illustrates the physical significance of theorem (5.90). It shows the case of N = 2, *i.e.*, a population containing two quasi-species. The set of the average eigenvalues makes up a two-dimensional area in the (v_1, v_2)-plane. Assume now that the system is at time t in the state (v_1', v_2'). Any virtual displacement $(\delta v_1, \delta v_2)$ must then lie on an ellipse, in order to be compatible with the constraints (5.91) and (5.92). The theorem (5.90) now states that any infinitesimal change in the v-variables will make the system take the path which corresponds to the greatest possible increase in the ordinate, that is, in the $\bar{\lambda}$-value.

5.6. Complementary Instruction and Selection in CP Systems

So far we have considered selection models in which each species directly "instructs" its own synthesis. In reality, however, the reproduction of a single-stranded nucleic acid is not a genuinely self-instructed process, since it requires a complementary intermediate (*cf.* Section 2.1). As we shall now show, selection in a system with complementary instruction is formally equivalent to genuine self-instruction, so that our reasoning above is still valid (Eigen, 1971; Thompson and McBride, 1974; Jones, 1976b).

In systems with cross-catalytic reproduction (or *complementary instruction*) each information unit is represented by a plus-minus-ensemble, that is, each unit of information now consists of a plus (+) and a minus (-) strand. Thus in systems with complementary instruction the phenomenological rate equations run

$$\dot{x}_{+i} = A_{+i}Q_{+i}x_{-i} - D_{+i}x_{+i} - \varphi x_{+i} + \sum_{\substack{\pm j \\ (j \neq -i)}} \psi_{+ij}x_j \qquad (i,j = 1,\ldots,N) \qquad (5.101)$$

where A_i, Q_i, D_i, φ and ψ_{ij} have their usual meanings. In order to keep the total population constant, the dilution parameter φ must be fitted to the average excess productivity such that

$$\varphi \equiv \bar{E} = \frac{1}{c}\sum_{i>0}(A_{-i}-D_{+i})x_{+i} + \frac{1}{c}\sum_{i>0}(A_{+i}-D_{-i})x_{-i} \qquad (5.102)$$

Rearranging Eq. (5.101) into the form of Eq. (5.39), we obtain the fundamental CP selection equations for systems with complementary instruction:

$$\dot{x}_i = \sum_j w_{ij} x_j - \bar{E}\, x_i \qquad (i,j = \pm 1,\ldots,\pm N) \tag{5.103}$$

The w_{ij} are the elements of the matrix

$$\mathbf{W} = (w_{ij}) = \begin{cases} A_i Q_i \delta_{-ij} - D_i\, \delta_{ij} + \psi_{ij}\delta_{ij} & \text{for } j = \pm i \\ \psi_{ij} & \text{for } j \neq \pm i \end{cases} \tag{5.104}$$

The set of equations (5.103) is formally identical to Eqs. (5.39); the only difference between them is that $\mathbf{W}$ is now a 2N x 2N matrix and not, as before, an N x N matrix, since each of the information carriers now consists of a pair of individuals inherently coupled by the plus-strand/minus-strand relationship.

Assuming that the elements of $\mathbf{W}$ are constant and that $\mathbf{W}$ does not have a degenerate spectrum of eigenvalues, the exact solution of the differential equations (5.103) is obtained in a way strictly analogous to the derivation of Eq. (5.63):

$$x_i(t) = \frac{c \sum_k u_{ik} a_k \exp(\lambda_k t)}{\sum_k \sum_i u_{ik} a_k \exp(\lambda_k t)} \qquad (i,k = \pm 1,\ldots,\pm N) \tag{5.105}$$

Here λ_k are the eigenvalues of the matrix $\mathbf{W}$ defined by Eq. (5.104), and u_{ik} are the components of the corresponding eigenvectors. The constants a_k are determined by the initial conditions (t=0):

$$a_j = \sum_{k>0} \left(u^{-1}_{j,+k}\, x_{+k}(0) + u^{-1}_{j,-k}\, x_{-k}(0) \right) \tag{5.106}$$

As Eq. (5.105) makes clear, the solutions depend critically upon the eigenvalue spectrum. Consider first of all the case where reverse mutations can be neglected ($\psi_{ij} \ll w_{ii}$). The off-diagonal elements of $\mathbf{W}$ are then all effectively zero and the eigenvalues are

$$\lambda_{\pm k} = -\frac{1}{2}(D_{+k}+D_{-k}) \pm \frac{1}{2}\sqrt{4(A_{+k}Q_{+k}A_{-k}Q_{-k} - D_{+k}D_{-k})+(D_{+k}+D_{-k})^2} \tag{5.107}$$

Each ($\pm$)-ensemble is characterised by two (real) eigenvalues. One eigenvalue, denoted λ_{-k}, is always negative. The complementary eigenvalue, denoted λ_{+k}, is positive if the growth rate exceeds the decay rate, *i.e.*,

$$A_{+k}Q_{+k}A_{-k}Q_{-k} > D_{+k}D_{-k} \tag{5.108}$$

This result is in principle unaltered if we include finite mutation terms ψ_{ij}, as long as these terms are small in comparison with w_{ii}. A second-order perturbation calculation in this case again provides useful approximate solutions for the eigenvalues and eigenvectors (*cf.* Eqs. 5.65 and 5.66).

Once the mutation terms ψ_{ij} become significant, they start to influence decisively the form of the eigenvalues. In particular, the eigenvalues can take on complex values, leading to oscillations in the concentration variables (see Jones, 1976b; Thompson and McBride, 1974). Let us now illustrate in more detail the physical aspects of cross-catalytic replication kinetics.

For simplicity we shall start with the unconstrained system (Eqs. 5.101 with $\varphi \equiv 0$) and neglect reverse mutations ($\psi_{ij} = 0$). This simplifies Eqs. (5.101) to a set of 2N differential equations coupled pairwise:

$$\begin{aligned} \dot{x}_{+i} &= A_{+i}Q_{+i}x_{-i} - D_{+i}x_{+i} \\ \dot{x}_{-i} &= A_{-i}Q_{-i}x_{+i} - D_{-i}x_{-i} \end{aligned} \qquad (i = 1,\ldots,N) \tag{5.109}$$

Each information unit now consists of two components and can be represented by a two-component column vector:

$$\mathbf{x}_i = \begin{pmatrix} x_{+i} \\ x_{-i} \end{pmatrix} \tag{5.110}$$

The selection value W_i^{CP} can further be represented by the matrix

$$\mathbf{W}_i = \begin{pmatrix} -D_{+i} & A_{+i}Q_{+i} \\ A_{-i}Q_{-i} & -D_{-i} \end{pmatrix} \tag{5.111}$$

whose eigenvalues λ_k are the solutions of the characteristic equation

$$\det(\mathbf{W}_i - \lambda \mathbf{I}) = 0 \tag{5.112}$$

Each eigenvalue is assigned to a normal mode of the reaction, and these are represented by the variables v_i. The v_i-variables are linear combinations of the x_{+i}- and x_{-i}-variables and satisfy the differential equations

$$\dot{v}_{\pm i} = \lambda_{\pm i} v_{\pm i} \qquad (i = 1,\ldots,N) \tag{5.113}$$

It should be emphasised that the assignment of "+" and "-" to the quantities v and λ is purely arbitrary, although for the concentration variable x the assignment can have biological meaning in that the coding strand is generally denoted "+".
The solutions of the characteristic equation (5.112) are (*cf.* Eq. 5.107)

$$\lambda_{\pm i} = -\frac{1}{2}(D_{+i}+D_{-i}) \pm \frac{1}{2}\sqrt{4(A_{+i}Q_{+i}A_{-i}Q_{-i})+(D_{+i}-D_{-i})^2} \tag{5.114}$$

We shall simplify further discussion by assuming that each strand of the $(\pm i)$-ensemble has the same decay rate parameter, *i.e.*,

$$D_{+i} = D_{-i} = D_i \tag{5.115}$$

With this approximation we have

$$\lambda_{\pm i} = \pm\sqrt{A_{+i}Q_{+i}A_{-i}Q_{-i}} - D_i \tag{5.116}$$

The physical interpretation of the eigenvalue analysis is clear: to each cross-catalytic reaction cycle two (real) eigenvalues are assigned which describe the time-dependent behaviour of the ensemble. One of these eigenvalues, which we arbitrarily call λ_{-i}, is always negative. It describes the process of equilibration between the plus- and minus-strands. To illustrate it we consider the corresponding normal mode v_{-i}, which we obtain as a solution of Eq. (5.113):

$$v_{-i} = v_{-i}(0)\ \exp\left[\left(-\sqrt{A_{+i}Q_{+i}A_{-i}Q_{-i}} - D_i\right)t\right] \tag{5.117}$$

Since the argument of the exponential function is negative, this normal mode decays. However, the v_i-variables are linear combinations of the x_{+i}- and x_{-i}-variables, which means that as $t \rightarrow \infty$ a constant ratio is set up between x_{+i} and x_{-i} (see Eq. 5.123).

The eigenvalue λ_{+i} is positive if

$$\sqrt{A_{+i}Q_{+i}A_{-i}Q_{-i}} > D_i \tag{5.118}$$

In this case the corresponding normal mode

$$v_{+i} = v_{+i}(0)\ \exp\left[\left(\sqrt{A_{+i}Q_{+i}A_{-i}Q_{-i}} - D_i\right)t\right] \tag{5.119}$$

represents the autocatalytic growth of the total ($\pm i$) ensemble.

If we write the two concentration variables $x_{\pm i}$ as the two components of a column vector (*cf.* Eq. 5.110) then the following vector relationship exists between the x- and the v-variables:

$$\mathbf{x}_i = \mathbf{U}_i\mathbf{v}_i \qquad \text{or} \qquad \mathbf{v}_i = \mathbf{U}_i^{-1}\mathbf{x}_i \tag{5.120}$$

where the matrix $\mathbf{U}$ and its inverse $\mathbf{U}^{-1}$ are determined by the components of the eigenvectors. A simple calculation (Appendix A.1) shows that

$$\mathbf{U}_i = \begin{pmatrix} 1 & -\sqrt{\dfrac{A_{+i}Q_{+i}}{A_{-i}Q_{-i}}} \\ \sqrt{\dfrac{A_{-i}Q_{-i}}{A_{+i}Q_{+i}}} & 1 \end{pmatrix} \tag{5.121}$$

i.e., the x- and v-variables are connected by the relations

$$x_{+i} = \frac{1}{2}\left(v_{-i} + \sqrt{\frac{A_{+i}Q_{+i}}{A_{-i}Q_{-i}}}\, v_{+i}\right) \tag{5.122}$$

and

$$x_{-i} = \frac{1}{2}\left(-\sqrt{\frac{A_{-i}Q_{-i}}{A_{+i}Q_{+i}}}\, v_{-i} + v_{+i}\right) \tag{5.123}$$

The equilibrium value of x_{+i}/x_{-i} follows from $v_{-i} \rightarrow 0$ for $t \rightarrow \infty$:

$$\lim_{t\to\infty} \frac{x_{+i}}{x_{-i}} = \sqrt{\frac{A_{+i}Q_{+i}}{A_{-i}Q_{-i}}} \tag{5.124}$$

For equal growth rate parameters of plus- and minus-strands the concentration ratio at equilibrium is unity. (*N.B.*: The $A_{+i}Q_{+i}$ are the parameters for formation of the plus-strand on the minus-strand template.)

If we now introduce selection pressure by applying the constraint of constant population, Eqs. (5.113), which describe the growth of normal modes, become

$$\dot{v}_{\pm i} = (\lambda_{\pm i} - \bar{E})v_{\pm i} \qquad (i = 1,\ldots,N) \tag{5.125}$$

These differential equations are formally identical to Eqs. (5.69), with the difference that $\bar{E}$ now satisfies the definition (5.102) of the average excess productivity in systems with complementary instruction.

None the less, Eqs. (5.125) show that the selection kinetics of CP systems with complementary instruction are described in principle by the same equations as are the selection kinetics of CP systems with genuine self-instruction. With complementary instruction the eigenvalues λ_i represent generalized selection values containing the geometric mean of the A_iQ_i parameters of the plus- and minus-strands (see Eq. 5.107). This is important, because the selective advantage from the geometric mean is enjoyed equally by the plus- and the minus-strands. As Eq. (5.124) shows, the worse template of a ($\pm$) ensemble always reaches a higher stationary-state concentration, which compensates for the inherently lower rate of synthesis of its complement. The more efficient a template is in building up a complementary strand, the lower will be its stationary-state concentration.

5.7. Selection Constraint CF: Constant Fluxes

We shall now investigate the kinetics of selection among macromolecules which are subject to the constraint of constant fluxes. We again consider the model system depicted in Fig. 4.1 but this time we shall adjust the reaction conditions so as to keep the inflow of energy-rich monomers constant.

Let us begin by formulating the rate equations for the concentrations of monomers and polymers in CF systems. We shall then attempt to solve these equations analytically for particular models.

Starting with the general phenomenological equations (5.10), we obtain for the rate of change of polymer concentration

$$\dot{x}_i = (F_i^* - D_i - \varphi)x_i + \sum_{j \neq i} \psi_{ij}^* x_j \qquad (i = 1,\ldots,N) \qquad (5.126)$$

For CP systems we have often assumed that in addition to the constant polymer concentration the energy-rich monomers are also buffered. In CF systems, however, this assumption no longer makes sense, even as an approximation. In order to express the dependence of the formation parameters F_i and ψ_{ij} upon the monomer concentrations, we shall from now on denote these parameters with an asterisk. We also define a set of *stoichiometric functions*

$$\sigma_{ij} = \sigma_{ij}(m_1,\ldots,m_\mu) \qquad (i,j = 1,\ldots,N) \tag{5.127}$$

such that the parameters F_i^* and ψ_{ij}^* are related to their values at fixed, standard monomer concentrations by

$$F_i^* = \sigma_{ii}(m_1,\ldots,m_\mu)F_i \tag{5.128}$$

and

$$\psi_{ij}^* = \sigma_{ij}(m_1,\ldots,m_\mu)\psi_{ij} \tag{5.129}$$

The exact form of the stoichiometric functions σ_{ij} naturally depends upon the particular mechanism involved in the polymerization reaction.

It will prove useful for our discussion to rewrite the phenomenological equations (5.126) in matrix notation. For this purpose we first condense all the parameters of formation into a matrix $\mathbf{F}$:

$$\mathbf{F} = (f_{ij}) = \begin{cases} F_i & \text{for } i = j \\ \psi_{ij} & \text{for } i \neq j \end{cases} \tag{5.130}$$

We then construct a matrix $\mathbf{D}$ from all the decay and dilution terms:

$$\mathbf{D} = (d_{ij}) = \begin{cases} D_i+\varphi & \text{for } i = j \\ 0 & \text{for } i \neq j \end{cases} \tag{5.131}$$

Finally we define a matrix $\mathbf{F}^*$ as

$$\mathbf{F}^* = (f_{ij}^*) = \begin{cases} F_i^* & \text{for } i = j \\ \psi_{ij}^* & \text{for } i \neq j \end{cases} \tag{5.132}$$

The elements of $\mathbf{F}^*$ thus depend upon the functions σ_{ij} (Eqs. 5.128 and 5.129), and therefore upon the monomer concentrations m_μ.

With these definitions we can express the set of differential equations (5.126) as

$$\dot{\mathbf{x}} = \left(\mathbf{F}^*(\mathbf{m}) - \mathbf{D}\right)\mathbf{x} \tag{5.133}$$

In CF systems the rate of change of the monomer concentrations is given by the following differential equations:

$$\dot{m} = \Phi_\mu - \sum_{i,j} \nu_{\mu i} f_{ij}^* x_j \qquad (\mu = 1,\ldots,4) \tag{5.134}$$

where $\nu_{\mu i}$ is the number of monomers of type μ in the sequence of the polymer I_i and Φ_μ is the rate of supply of monomer μ. Equations (5.134) are only valid in the form given when the efflux of energy-rich monomers that have not been incorporated is negligible. Let us express all Φ_μ as a vector $\mathbf{\Phi}$, all m_μ as a vector $\mathbf{m}$ and all $\nu_{\mu i}$ as a matrix $\mathbf{N}$. Equations (5.134) may then be rewritten as

$$\dot{\mathbf{m}} = \mathbf{\Phi} - \mathbf{N}\mathbf{F}^*(\mathbf{m})\mathbf{x} \qquad (\mathbf{\Phi} = \text{const.}) \tag{5.135}$$

The CF constraint requires that the flux $\mathbf{\Phi}$ which transports the energy-rich monomers into the system from outside be constant in time (*cf.* Eq. 5.14).

Equations (5.133) and (5.135) are the rate equations for the concentrations of monomers and polymers in CF systems. We shall often refer to them as the *CF selection equations*.

5.8. Solution of the CF Selection Equations

The phenomenological equations for the CF system have a complex structure. We shall therefore have to restrict our discussion of Eqs. (5.133) and (5.135) to cases where simplifying assumptions can be made. Our treatment will be based on the work of I.R. Epstein and M. Eigen (1979).

First we shall analyse the stationary states of the CF system. We obtain the steady-state solutions by setting the left-hand sides of Eqs. (5.133) and (5.135) to zero. From Eq. (5.135) we first obtain the stationary-state solutions for the polymer concentrations, *i.e.*,

$$\mathbf{x}^s = \mathbf{F}^{*-1}(\mathbf{m}^s)\mathbf{N}^{-1}\mathbf{\Phi} \tag{5.136}$$

Insertion of Eq. (5.136) into Eq. (5.133) and rearrangement lead to an equation from which the stationary-state monomer concentration can be obtained:

$$(\mathbf{D}^{-1}\mathbf{N}^{-1} - \mathbf{F}^{*-1}\mathbf{N}^{-1})\mathbf{\Phi} = 0 \tag{5.137}$$

If the dependence of the stoichiometric functions $\sigma_{ij}(\mathbf{m})$ upon the monomer concentrations is known, we can in principle calculate the steady-state concentrations both of the monomers and of the macromolecular species from Eqs. (5.136) and (5.137). This will be done for a simple case in Section 5.10.

Apart from the complex structure of the selection equations (5.133) and (5.135) it is above all our ignorance of the form of the stoichiometric functions σ_{ij} which prevents a full analytical solution of the CF system. However, exact solutions can be given if we make a number of simplifying approximations.

FIRST APPROXIMATION. The first approximation sets out from the assumption that the decay parameters D_i for all species I_i are either equal or much smaller than the dilution parameter φ:

$$D_i + \varphi \approx D \qquad (i = 1,\ldots,N) \tag{5.138}$$

This approximation leads to

$$\mathbf{D} = D\,\mathbf{I} \tag{5.139}$$

i.e. the matrix $\mathbf{D}$ is equal to the identity matrix $\mathbf{I}$ multiplied by the constant D.

SECOND APPROXIMATION. Under natural conditions all molecular species in a self-organizing genetic system are coupled to one another by the mutation terms ψ_{ij}. As selection equilibrium is approached the polymer distribution will correspond to a quasi-species consisting of a master sequence I_m and its accompanying mutant distribution. Since under natural conditions the reproduction accuracy Q_m of the master sequence should not be far from unity, all the mutant sequences will closely resemble the master sequence and therefore each other as well. It is therefore a realistic approximation if we replace the stoichiometric functions of the individual species $\sigma_{ij}(\mathbf{m})$ by an average function $\sigma(\mathbf{m})$ for all species:

$$\sigma_{ij}(\mathbf{m}) \approx \sigma(\mathbf{m}) \qquad (i,j = 1,\ldots,N) \tag{5.140}$$

If we now replace the vector $\mathbf{m}$ by a scalar m (with a scalar flux Φ_m), and describe the monomer composition $\nu_{\mu i}$ by the single quantity ν, the result is a considerable simplification of the CF selection equations. Equations (5.133) and (5.135) become respectively

$$\dot{x}_i = \sigma(m) \sum_j f_{ij} x_j - D x_i \qquad (i = 1,\ldots,N) \tag{5.141}$$

and

$$\dot{m} = \Phi_m - \nu\sigma(m) \sum_{i,j} f_{ij}x_j \, . \tag{5.142}$$

In CF systems we have the conservation condition (*cf.* Eq. 5.9)

$$\sum_{i,j} f_{ij}x_j = \sum_j A_j x_j \tag{5.143}$$

which in turn allows Eq. (5.142) to be rewritten as

$$\dot{m} = \Phi_m - \nu\sigma(m) \sum_j A_j x_j \tag{5.144}$$

Equations (5.141) and (5.144) make up a set of simplified CF selection equations for which exact solutions can now be constructed. This is done in a way very similar to the procedure described in Section 5.3.

We first define a new set of variables q_i by

$$x_i = h(t)\, q_i(t) \qquad (i = 1,\ldots,N) \tag{5.145}$$

where

$$h(t) = \exp(-Dt) \tag{5.146}$$

Insertion of Eqs. (5.145) into Eqs. (5.141) gives a set of N linear differential equations

$$\dot{q}_i = \sigma(m) \sum_j f_{ij}q_j \qquad (i = 1,\ldots,N) \tag{5.147}$$

or, in vector notation,

$$\dot{\mathbf{q}} = \sigma(m)\mathbf{F}\mathbf{q} \tag{5.148}$$

We now treat Eq. (5.148) as an eigenvalue problem (see Appendix A.1). Proceeding in the standard way, we consider the matrix $\mathbf{U} = (u_{ij})$ of the eigenvectors of $\mathbf{F}$ and the corresponding eigenvalue matrix $\mathbf{L}$:

$$\mathbf{L} = (l_{ij}) = \begin{cases} \lambda_i & \text{for } i = j \\ 0 & \text{for } i \neq j \end{cases} \tag{5.149}$$

The following equivalent relations then hold (written respectively in vector and scalar notation)

$$\mathbf{U}^{-1}\mathbf{F}\,\mathbf{U} = \mathbf{L} \tag{5.150}$$

or

$$\sum_j f_{ij}u_{jk} = \lambda_k u_{ik} \qquad (i = 1,\ldots,N) \tag{5.151}$$

We now define a further vector $\mathbf{v}$ by

$$\mathbf{q} = \mathbf{U}\mathbf{v} \tag{5.152}$$

Substituting Eq. (5.152) into Eq. (5.148) gives

$$\mathbf{U}\dot{\mathbf{v}} = \sigma(m)\mathbf{F}\mathbf{U}\mathbf{v} \tag{5.153}$$

which multiplied by $\mathbf{U}^{-1}$ and combined with Eq. (5.150) leads to

$$\dot{\mathbf{v}} = \sigma(m)\mathbf{L}\mathbf{v} \tag{5.154}$$

or, in scalar notation,

$$\dot{v}_i = \sigma(m)\lambda_i v_i \qquad (i = 1,\ldots,N) \tag{5.155}$$

The v_i-variables are again the normal modes of reaction (*cf.* Eq. 5.46). Since σ, as a function of monomer concentrations, is implicitly time-dependent, we cannot solve Eqs. (5.155) without first solving Eq. (5.142). We can however determine the trajectories of the system in the (v_i, v_j)-plane. For this purpose we examine the quotients

$$\frac{\dot{v}_i}{\dot{v}_j} = \frac{\lambda_i v_i}{\lambda_j v_j} \qquad (i,j = 1,\ldots,N) \tag{5.156}$$

in which the stoichiometric function $\sigma(m)$ no longer appears. Integration of Eq. (5.156) yields

$$\frac{v_i}{v_i(0)} = \left(\frac{v_j}{v_j(0)}\right)^{\frac{\lambda_i}{\lambda_j}} \qquad (i,j = 1,\ldots,N) \tag{5.157}$$

With the aid of Eqs. (5.157) all the v_i values can be determined as long as the initial conditions $\mathbf{v}(0)$ (or $\mathbf{x}(0)$) and the value of any of the quantities v_j at time t are known.

As already mentioned, complete solutions of Eqs. (5.155) can only be given when we can also solve Eq. (5.142). To do this, we introduce the variables v_k into Eq. (5.142). Combining this with Eqs. (5.145), (5.146), (5.151) and (5.152) leads to the result

$$\dot{m} = \Phi_m - v\sigma(m) \sum_{i,k} \lambda_k u_{ik} v_k \exp(-Dt) \tag{5.158}$$

Following Eq. (5.155) we can now express the function σ in terms of one of the variables v_i. We can, without restricting the generality of our treatment, choose the variable v_m which corresponds (by definition) to the highest eigenvalue λ_m. This substitution leads finally from Eq. (5.158) to

$$\dot{m} = \Phi_m - v\frac{\dot{v}_m}{\lambda_m v_m} \sum_{i,k} \lambda_k u_{ik} v_k \exp(-Dt) \tag{5.159}$$

In order to solve this differential equation analytically we must make one further assumption.

THIRD APPROXIMATION. We assume that the concentration of energy-rich monomers is close to its stationary-state value:

$$\dot{m} \approx 0 \tag{5.160}$$

This condition is fulfilled to a good approximation when the concentration of free, energy-rich monomers adapts itself to changes in the flux Φ_m on a time-scale much shorter than that of the *selective* change of polymer concentrations, so that for the polymers the monomer concentration m appears constant.
From Eq. (5.144) and approximation (5.160) we obtain

$$\Phi_m = v\sigma(m) \sum_j A_j x_j \tag{5.161}$$

Equations (5.141) further yield, summing over all i, the relation

$$\sum_i \dot{x}_i = \sigma(m) \sum_{i,j} f_{ij} x_j - D \sum_i x_i \tag{5.162}$$

Combining this equation with the conservation relationship (5.143) and the relation (5.161) we obtain

$$\sum_i \dot{x}_i = \frac{\Phi_m}{v} - D \sum_i x_i \tag{5.163}$$

We now denote the total concentration of all species I_i as c and obtain finally for its time-dependence under CF constraint

$$\dot{c}(t) = \frac{\Phi_m}{v} - Dc(t) \tag{5.164}$$

This equation has for $\dot{c}=0$ the stationary-state solution

$$c^S = \frac{\Phi_m}{\nu D} \tag{5.165}$$

The full solution of Eq. (5.164) is

$$c(t) = c^S + \{c(0)-c^S\} \exp(-Dt) \tag{5.166}$$

Consider now the quantity v_m, corresponding to the species with the highest eigenvalue λ_m. This satisfies, in accordance with Eq. (5.155), the differential equation

$$\sigma = \frac{\dot{v}_m}{\lambda_m v_m} \tag{5.167}$$

Note that Eq. (5.157) allows the expression of every v_i as a function of the initial conditions $\mathbf{v}(0)$ and of any v_j. In particular, we have for v_m

$$v_i = v_i(0) \left(\frac{v_m}{v_m(0)} \right)^{\lambda_i/\lambda_m} \tag{5.168}$$

If we substitute the relation (5.157) into Eq. (5.159) we obtain

$$\dot{m} = \Phi_m - \nu\exp(-Dt) \sum_{i,k} u_{ik}(a_k+1) \frac{v_k(0)}{v_m(0)^{a_k+1}} v_m^{a_k} \dot{v}_m \tag{5.169}$$

where the constants a_k are defined by

$$a_k = \frac{\lambda_k}{\lambda_m} - 1 \tag{5.170}$$

We now change variables, with

$$\theta_k = v_m^{a_k+1} \tag{5.171}$$

Differentiation of Eq. (5.171) gives the following equation for θ_k:

$$\dot{\theta}_k = (a_k+1)\, v_m^{a_k}\, \dot{v}_m \tag{5.172}$$

Substitution of Eq. (5.172) into Eq. (5.169) yields

$$\dot{m} = \Phi_m - \nu\exp(-Dt) \sum_{i,k} u_{ik} \frac{v_k(0)}{v_m(0)^{a_k+1}} \dot{\theta}_k \tag{5.173}$$

If we now assume a quasi-stationary state for the monomer concentration, we can make use of the approximation (5.160). We thus set the left-hand side Eq. (5.173) to zero and integrate directly with respect to t. After simple rearrangement we obtain

$$\sum_{i,k} u_{ik} v_k(0) \left(\frac{v_m}{v_m(0)}\right)^{\lambda_k/\lambda_m} = \frac{\Phi_m}{\nu D} \{\exp(Dt)-1\} + \sum_{i,k} u_{ik} v_k(0) \tag{5.174}$$

Equations (5.145) and (5.152) imply that at time t=0

$$\sum_{i,k} u_{ik} v_k(0) = \sum_i q_i(0) = \sum_i x_i(0) = c(0) \tag{5.175}$$

Substituting Eqs. (5.165) and (5.175) into Eq. (5.174), we obtain finally

$$\sum_{i,k} u_{ik} v_k(0) \left(\frac{v_m}{v_m(0)}\right)^{\lambda_k/\lambda_m} = c^S \exp(Dt) - c^S + c(0) \tag{5.176}$$

This is the complete, time-dependent solution for the variable $v_m = v_m(t)$. All other normal modes $v_{i\neq m}(t)$ can then be calculated from Eq. (5.157).

5.9. Selection Dynamics in CF Systems

We shall now examine the properties of the solution (5.176) of the CF selection equations.

Let us first consider the asymptotic behaviour at $t \to \infty$. Since $\lambda_j < \lambda_m$ (for $j \neq m$), we have

$$v_m(t) \sim \exp(Dt) \tag{5.177}$$

and from Eq. (5.157)

$$v_j(t) \sim \exp\left(\frac{\lambda_j Dt}{\lambda_m}\right) \tag{5.178}$$

The quotient v_j/v_m shows for $t \to \infty$ the asymptotic behaviour

$$\frac{v_j}{v_m} \sim \exp\left\{\left(\frac{\lambda_j}{\lambda_m} - 1\right) Dt\right\} \to 0 \qquad (j \neq m) \tag{5.179}$$

The kinetics of selection expressed by this equation are qualitatively the same as those of a quasi-species in a CP system, in so far as in selection equilibrium ($t \to \infty$) only the quasi-species with the highest eigenvalue λ_m survives (*cf.* Section 5.4).

Since the concentration of each molecular species I_i can in accordance with Eqs. (5.145) and (5.152) be represented as a linear combination of the v_j-variables,

$$x_i(t) = \sum_j u_{ij} v_j(t) \exp(-Dt) \tag{5.180}$$

then for every species I_i we have in selection equilibrium, as a direct consequence of Eq. (5.179),

$$\lim_{t \to \infty} x_i(t) = u_{im} v_m(t) \exp(-Dt) \tag{5.181}$$

From this follows for $c(t\to\infty) = c^S$ the relation

$$\lim_{t\to\infty} v_m(t) = \frac{c^S \exp(Dt)}{\sum_i u_{im}} \tag{5.182}$$

If Eqs. (5.181) and (5.182) are combined, the following limiting behaviour can be deduced for species I_i:

$$\lim_{t\to\infty} \frac{x_i(t)}{c^S} = \frac{u_{im}}{\sum_i u_{im}} \qquad (i = 1,\ldots,N) \tag{5.183}$$

Thus in selection equilibrium the relative concentration of each species I_i is equal to its relative contribution to the preponderant quasi-species. This is the same result as was obtained for selection in CP systems (*cf.* Eq. 5.84).

If $f_{ij} \ll f_{ii}$ $(i \neq j)$ then a perturbation calculation leads to the approximate solutions for the eigenvalues (*cf.* Eq. 5.65)

$$\lambda_i \approx f_{ij} + \sum_{j\neq i} \frac{f_{ij}\, f_{ji}}{f_{ii}-f_{jj}} = A_iQ_i + \sum_{j\neq i} \frac{\psi_{ij}\,\psi_{ji}}{A_iQ_i-A_jQ_j} \tag{5.184}$$

and for the components of the corresponding eigenvectors (*cf.* Eq. 5.66)

$$u_{ij} \approx \delta_{ij} + \frac{f_{ij}}{f_{ij}-f_{ii}} = \delta_{ij}\,\frac{\psi_{ij}}{A_jQ_j-A_iQ_i} \tag{5.185}$$

Since all $f_{ij} = \psi_{ij}$ are non-negative the components u_{im} of the eigenvector corresponding to the dominant quasi-species are also non-negative, so that the asymptotic solutions (5.183) are also physically meaningful. The remarks made at the end of Section 5.3 concerning the general properties of the eigenvectors are also applicable here.

Finally, we shall take a look at an example of selection kinetics in CF systems, and compare the selection behaviour under CP and CF constraints. For this purpose we now consider competition between two molecular species I_1 and I_2. The relevant parameters for our model will be:

$$A_1 = 5 \ , \ Q_1 = 0.99 \ , \ D_1 = D = 1 \ , \ f_{12} = 0.02$$
$$A_2 = 2 \ , \ Q_2 = 0.99 \ , \ D_2 = D = 1 \ , \ f_{21} = 0.05 \qquad (5.186)$$
$$\Phi_m = c(0) = c^s = 10 \ , \ \sigma(m) = 1$$

The eigenvalues and eigenvectors of the matrix **F** (see Eqs. 5.150 ff.) are then in a CF system

$$\mathbf{L}^{CF} = (l_{ij}) = \begin{pmatrix} 4.95034 & 0 \\ 0 & 1.97966 \end{pmatrix} \qquad (5.187)$$

and

$$\mathbf{U}^{CF} = (u_{ij}) = \begin{pmatrix} 0.983444 & -0.006736 \\ 0.016556 & 1.006736 \end{pmatrix} \qquad (5.188)$$

The second matrix has been normalized so that the condition

$$\sum_{i=1}^{2} u_{ij} = 1 \qquad (j = 1,2) \qquad (5.189)$$

is satisfied.

We now consider Eq. (5.166) and insert values for the parameters D and Φ_m. Since we have chosen a value for c(0) equal to c^s, the total concentration is now constant, *i.e.*,

$$c(t) = x_1(t) + x_2(t) = c(0) \qquad (5.190)$$

This means that, in this case, the population of the CF system remains constant, so that a direct comparison of the dynamics of selection under CF and CP conditions is possible.

If we now calculate for comparison the eigenvalues and eigenvectors of the matrix **W** (see Eq. 5.59), we obtain

$$\mathbf{L}^{CP} = (l_{ij}) = \begin{pmatrix} 3.950337 & 0 \\ 0 & 0.979663 \end{pmatrix} \tag{5.191}$$

and

$$\mathbf{U}^{CP} = (u_{ij}) = \begin{pmatrix} 0.983444 & -0.006736 \\ 0.016556 & 1.006736 \end{pmatrix} \tag{5.192}$$

Comparing the ratios of the eigenvalues for the CP case with those for the CF case, we have respectively

$$\left(\frac{\lambda_1}{\lambda_2}\right)^{CP} = 4.03 \tag{5.193}$$

and

$$\left(\frac{\lambda_1}{\lambda_2}\right)^{CF} = 2.50 \tag{5.194}$$

Thus under CP conditions the ratio of the eigenvalues is considerably higher than under CF conditions. This is easy to see, as in the CP system the selection value is given by the difference between AQ and D (see Definition 5.17) whereas in the CF system it is their quotient (Definition 5.203). The ratio of selection values is determined by the parameter values in Eqs. (5.186):

$$\frac{W^{CP}}{W^{CF}} = \begin{cases} 0.79 & \text{for species } I_1 \\ 0.49 & \text{for species } I_2 \end{cases} \tag{5.195}$$

The equilibrium values for x_1 and x_2 are, however, identical in the CP and CF systems, since the stationary-state values for the species I_i are given solely by the components of the eigenvector (Eqs. 5.84 and 5.183 respectively) belonging to the highest eigenvalue. The eigenvector matrices $\mathbf{U}^{CF}$ and $\mathbf{U}^{CP}$ are in this case identical.

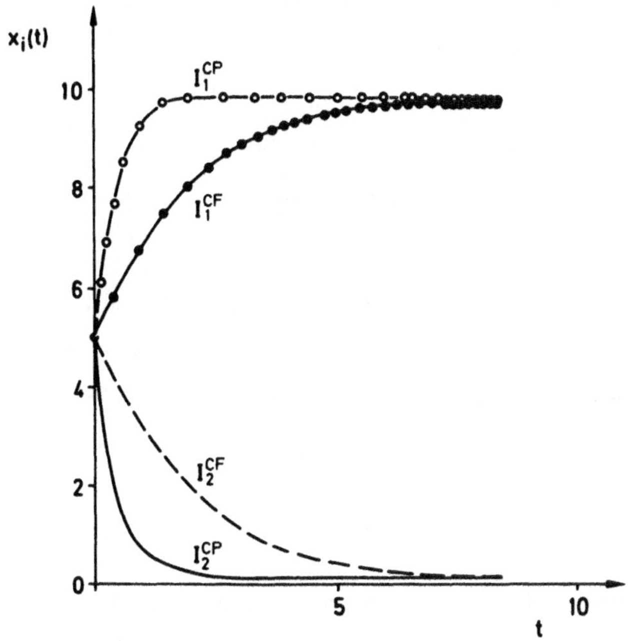

Fig. 5.5. Competition between two species I_1 and I_2 under CP and CF constraints. The initial conditions are the same for both species: $x_1(0) = x_2(0)$. In CP systems the selection is sharper than in CF systems, *i.e.*, selection equilibrium is reached more quickly in a CP system, even though the final distribution is the same in each: $x_1(\infty) = 9.83444$, $x_2(\infty) = 0.16556$.
The particular choice of values of the kinetic parameters (see Eq. 5.186) results in a constant total concentration of species, so that the CF system is effectively under CP constraint. Consequently the CF system has fewer degrees of freedom for reaching selection equilibrium than the CP system. (After Epstein and Eigen, 1979.)

Figure 5.5 shows a computer simulation of selection competition between the two species I_1 and I_2 under CP and CF constraints. We notice an important difference: the relation

$$\left(\frac{\lambda_1}{\lambda_2}\right)^{CP} > \left(\frac{\lambda_1}{\lambda_2}\right)^{CF} \tag{5.196}$$

leads to more rapid selection in the CP system. This difference in selection kinetics between CP and CF systems is fundamental and lies in the different structures of the respective selection values. Under CP conditions a self-organizing system can adapt itself more rapidly to changes in the environment than it can under

CF conditions. The concentration distributions in selection equilibrium are however identical under CP and CF constraints.

5.10. Competition and Co-existence in CF Systems

We shall now examine some further properties of selection in CF systems, again making use of simple models. One property, mutual exclusion by competition, is already familiar. The other, co-existence, is not shared by CP systems. To show the possibility of co-existence in CF systems we shall simplify matters by assuming that:

(a) All polymers have approximately the same monomer composition, so that we can replace the matrix $\mathbf{N} = (\nu_{\mu i})$ with the scalar quantity ν.

(b) All energy-rich monomers flow into the system at the same speed, so that we can replace the vector $\mathbf{\Phi} = \{\Phi_\mu\}$ with the scalar Φ_m.

(c) For every species I_i the error rate in reproduction is negligible ($Q_i \approx 1$, $\psi_{ij} \approx 0$).

(d) There is no dilution in the system ($\Phi_0 \equiv 0$).

(e) The kinetic parameters A_i and D_i are independent of the concentration variables x_k.

The monomer concentrations m_μ can be replaced by an average variable m and the approximation (5.140) again applies, that is,

$$\sigma_{ij}(\mathbf{m}) \approx \sigma(\mathbf{m}) = \sigma(m) \tag{5.197}$$

With assumptions (a) - (e) the phenomenological rate equations for the polymer concentrations become (*cf.* Eq. 5.133):

$$\dot{x}_i = \{\sigma(m)A_i - D_i\}x_i \tag{5.198}$$

The non-linear coupling of these equations is a consequence of the stoichiometric function $\sigma(m)$, since $m = m(x_1,\ldots\ldots,x_N)$ depends in general on all concentration variables x_k.

In order to solve Eqs. (5.198) we need a further differential equation to describe the monomer concentration as a function of the x_k. For this we recourse to Eq. (5.144) and obtain

$$\dot{m} = \Phi_m - \nu\sigma(m)\sum_k A_k x_k \tag{5.199}$$

As pointed out in the foregoing section, the solutions of Eqs. (5.198) and (5.199) depend critically on the form of the stoichiometric function $\sigma(m)$.
For subsequent discussion let us assume that $\sigma(m)$ has the simple form

$$\sigma(m) = m \tag{5.200}$$

and then analyse two different CF systems.

(1) COMPETITION. For $\sigma(m) = m$ the differential equations (5.198) and (5.199) have the following respective stationary-state solutions:

$$\mathbf{x}^{s_1} = \begin{pmatrix} \frac{\Phi_m}{\nu D_1} \\ 0 \\ \vdots \\ 0 \end{pmatrix}, \ldots\ldots, \mathbf{x}^{s_i} = \begin{pmatrix} 0 \\ \vdots \\ 0 \\ \frac{\Phi_m}{\nu D_i} \\ 0 \\ \vdots \\ 0 \end{pmatrix}, \ldots\ldots, \mathbf{x}^{s_N} = \begin{pmatrix} 0 \\ \vdots \\ 0 \\ \frac{\Phi_m}{\nu D_N} \end{pmatrix} \tag{5.201}$$

and

$$\mathbf{m}^{s_1} = \begin{pmatrix} \frac{D_1}{A_1} \\ 0 \\ \vdots \\ \vdots \\ 0 \end{pmatrix}, \ldots\ldots, \mathbf{m}^{s_i} = \begin{pmatrix} 0 \\ \vdots \\ 0 \\ \frac{D_i}{A_i} \\ 0 \\ \vdots \\ 0 \end{pmatrix}, \ldots\ldots, \mathbf{m}^{s_N} = \begin{pmatrix} 0 \\ \vdots \\ \vdots \\ \vdots \\ \frac{D_N}{A_N} \end{pmatrix} \tag{5.202}$$

The steady-state values are obtained by setting the left-hand sides of Eqs. (5.198) and (5.199) to zero. We first calculate $\mathbf{x}^{s_i}$ from Eq. (5.198) and then find $\mathbf{m}^{s_i}$ by inserting the steady-state concentration into Eq. (5.199).

It will prove useful to define the quantity

$$W_i^{CF} = \frac{A_i Q_i}{D_i} \tag{5.203}$$

as the selection value of a species I_i under CF constraints. Linear stability analysis (Appendix A.2) for Eqs. (5.198) and (5.199) shows that, of all possible stationary states, the stable one is that for which

$$W_m^{CF} = \max_i W_i^{CF} \tag{5.204}$$

applies, *i.e.*, the species with the highest selection value W_m^{CF} is the ultimate winner in the selection competition, while all other species die out. Definition (5.203) thus provides us with a formal analogy to the selection value in CP systems.

(2) CO-EXISTENCE. In this model, first discussed by J. Tyson (1974), we start with the same kinetic presuppositions as in case (1). This time however we make the additional assumption that each molecular species I_i is encapsulated in a primitive membrane (Fig. 5.6). The energy-rich monomers must therefore first diffuse through a membrane before becoming available to the molecular species I_i.

The kinetic equations for the polymer concentrations x_i are in principle the same as in case (1). However we must now specify the equations for the monomers rather more precisely.

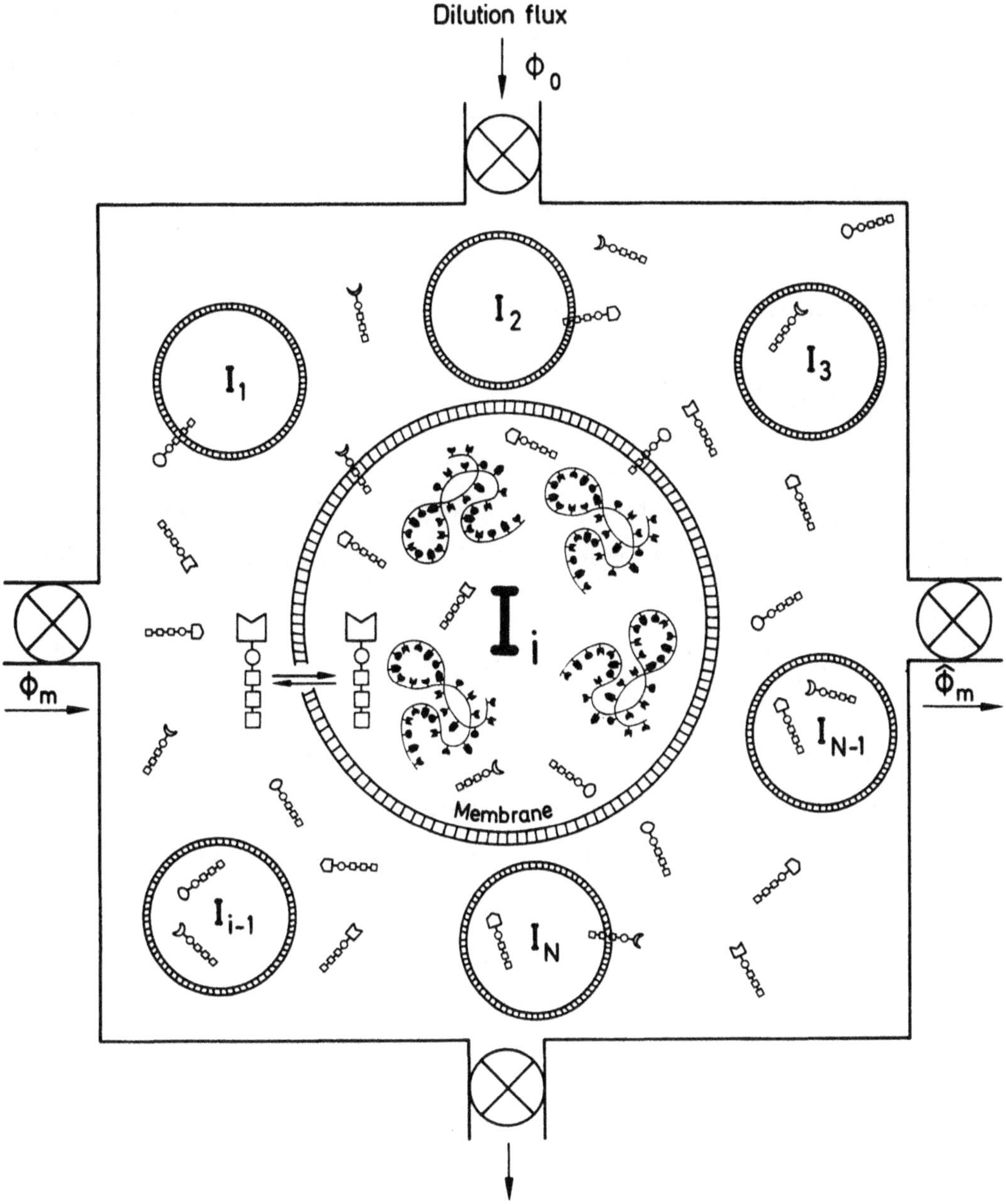

Fig. 5.6. Model for a compartmented CF system. Each class of species I_i is encapsulated in a membrane. The energy-rich monomers must diffuse through the membrane before new polymers can be synthesized. The concentration of monomers in the ith compartment is denoted $m^{(i)}$. The other constraints are the same as given in Fig. 4.1.

Let us denote the concentration of the energy-rich monomers within the ith compartment as $m^{(i)}$. From assumption (5.200) the stoichiometric function has the form $\sigma^{(i)} = m^{(i)}$. The concentration of species I_i in its compartment is then given by

$$\dot{x}_i = (m^{(i)}A_i - D_i)\, x_i \qquad (i = 1,\ldots,N) \tag{5.205}$$

while the concentration of energy-rich monomers in the ith compartment is given by

$$\dot{m}^{(i)} = \mathcal{D}_{eff}(m - m^{(i)}) - \nu m^{(i)} A_i x_i \tag{5.206}$$

in which $\mathcal{D}_{eff}$ is the effective *diffusion coefficient* for the monomers through the compartment walls. The rate of change in the overall concentration m is described by

$$\dot{m} = \Phi_m - \mathcal{D}_{eff} \sum_i (m - m^{(i)}) \tag{5.207}$$

There are numerous stationary states for the above differential equations. They are given without proof and classified in Table 5.1. As we see, the stability of the stationary states depends upon the energy input Φ_m. Thus the state S_i exists for all values for Φ_m. Other stationary states exist only when the energy input exceeds a critical level Φ_m^{crit}. We obtain the critical value of $\Phi_m^{(i)}$ by evaluating the steady-state concentrations of x_i (*cf.* Table 5.1) for $x_i > 0$.

This is not the place for a lengthy analysis of stability. Instead we shall make use of a plausibility argument, due to J. Tyson (1974), about the nature of the stationary-state solutions.

For this purpose we number the selection values of the N species so that they fall into order as given by the inequality

$$W_1^{CF} \geq W_2^{CF} \geq W_3^{CF} \geq \ldots\ldots \geq W_N^{CF} \tag{5.208}$$

Region of stability	Stationary polymer concentration	Stationary monomer concentration	Range of subscripts
S_1	$x_1 = \frac{1}{\nu A_1} W_1^{CF} \Phi_m$	$m^{(1)} = \frac{1}{W_1^{CF}}$	
	$x_j = 0$	$m^{(j)} = m = \frac{\Phi_m}{\mathcal{D}_{eff}} + \frac{1}{W_1^{CF}}$	for all $j \neq 1$
$S_{1,2}$	$x_1 = \frac{1}{2\nu A_1} \left[W_1^{CF} \Phi_m - \mathcal{D}_{eff} \left(1 - \frac{W_1^{CF}}{W_2^{CF}}\right) \right]$	$m^{(1)} = \frac{1}{W_1^{CF}}$	
	$x_2 = \frac{1}{2\nu A_2} \left[W_2^{CF} \Phi_m - \mathcal{D}_{eff} \left(1 - \frac{W_2^{CF}}{W_1^{CF}}\right) \right]$	$m^{(2)} = \frac{1}{W_2^{CF}}$	
	$x_j = 0$	$m^{(j)} = m = \frac{1}{2} \left(\frac{\Phi_m}{\mathcal{D}_{eff}} + \frac{1}{W_1^{CF}} + \frac{1}{W_2^{CF}} \right)$	for all $j \neq 1,2$
$\vdots$	$\vdots$	$\vdots$	
$S_{1,\ldots,N}$	$x_1 = \frac{1}{N\nu A_1} \left[W_1^{CF} \Phi_m - \mathcal{D}_{eff} \left(N-1- \sum_{i=2}^{N} \frac{W_1^{CF}}{W_i^{CF}}\right) \right]$	$m^{(1)} = \frac{1}{W_1^{CF}}$	
	$\vdots$	$\vdots$	
	$x_N = \frac{1}{N\nu A_N} \left[W_N^{CF} \Phi_m - \mathcal{D}_{eff} \left(N-1- \sum_{i=1}^{N-1} \frac{W_N^{CF}}{W_i^{CF}}\right) \right]$	$m^{(N)} = \frac{1}{W_N^{CF}}$	
		$m = \frac{1}{N} \left(\frac{\Phi_m}{\mathcal{D}_{eff}} + \sum_{i=1}^{N} \frac{1}{W_i^{CF}} \right)$	

Table 5.1. Classification of the stationary-state solutions of the sets of differential equations (5.205), (5.206), (5.207)

The stable stationary states then appear, along with their regions of stability, in the order shown in Table 5.2.

In our model we switch on the energy input flux Φ_m at $t = t_0$. For small values of Φ_m only one molecular species, the one with the highest selection value W_1^{CF}, can exist in the long term. But as soon as the energy input exceeds a threshold rate $\Phi_m^{(2)}$, a second species I_2 can co-exist with the first species; this is the one with the second highest selection value W_2^{CF}. If the energy flux is increased further, a critical value $\Phi_m^{(3)}$ is reached where a third species with the next highest selection value W_3^{CF} can co-exist, and so forth. If $\Phi_m \geq \Phi_m^{(N)}$ then all the species I_i can co-exist, whereby their concentrations, in accordance with the inequality (5.208), are related by

$$x_1^{s_N} \geqq x_2^{s_N} \geqq x_3^{s_N} \geqq \ldots\ldots \geqq x_N^{s_N} > 0 \tag{5.209}$$

The magnitude of the selection pressure in this model obviously depends upon the energy input flux Φ_m. The lower this is, the more rigorous the selection; the higher the energy flux, the broader the spectrum of co-existing species.

Thus in contrast to selection in homogeneous CF systems, selection in a compartmented system results under certain conditions (*i.e.*, a sufficiently high energy input rate) in the co-existence of various species. The advantages of such a mechanism for the evolutionary development of such systems is clear. In the homogeneous case, the optimization process depends exclusively upon the chance appearance of a selectively advantageous mutant. Evolution is then a process which makes random decisions at branching-points and can reach quiescence at a local optimum which lies far below the global optimum. If compartmentation is present several species can co-exist, so that each compartment becomes a separate centre of evolution. Thus in the latter case several evolutionary paths are tried out at once. Since a short-term increase in monomer supply widens the range of species which can co-exist, it is even possible for the average selection value to decrease and thus to pass through local minima.

Region of stability	Critical energy flow	Selection equilibrium
S_1	$\Phi_m^{(1)} > 0$	$x_1 > 0$
$S_{1,2}$	$\Phi_m^{(2)} > \mathcal{D}_{eff}\left(\frac{1}{W_2^{CF}} - \frac{1}{W_1^{CF}}\right)$	$x_1 \geqq x_2 > 0$
$S_{1,2,3}$	$\Phi_m^{(3)} > \mathcal{D}_{eff}\left(\frac{2}{W_3^{CF}} - \frac{1}{W_2^{CF}} - \frac{1}{W_1^{CF}}\right)$	$x_1 \geqq x_2 \geqq x_3 > 0$
$S_{1,2,3,4}$	$\Phi_m^{(4)} > \mathcal{D}_{eff}\left(\frac{3}{W_4^{CF}} - \frac{1}{W_3^{CF}} - \frac{1}{W_2^{CF}} - \frac{1}{W_1^{CF}}\right)$	$x_1 \geqq x_2 \geqq x_3 \geqq x_4 > 0$
⋮	⋮	⋮
$S_{1,\ldots\ldots,N}$	$\Phi_m^{(N)} > \mathcal{D}_{eff}\left(\frac{N-1}{W_N^{CF}} - \sum_{i=1}^{N-1} \frac{1}{W_i^{CF}}\right)$	$x_1 \geqq x_2 \geqq \ldots\ldots \geqq x_N > 0$

Table 5.2. Regions of co-existence in the compartmented CF system (for details see text).

All in all a compartmented system is clearly more versatile in its selection properties, since at low energy input rates Φ_m the system simulates the homogeneous case of rigorous competition. At high energy fluxes Φ_m the system allows the co-existence of different species. If large fluctuations in Φ_m should occur, the phases of co-existence and unambiguous selection will alternate, so that evolution will proceed more rapidly and completely.

6. Stochastic Theory of Selection

Up to now our discussion has shown that mutations are the source of evolutionary change and progress. However, the elementary process which leads to a *specific* mutation is fundamentally undetermined. Consequently, deterministic selection theory is not in a position to predict the moment of appearance or the nature of a mutation. It is concerned rather with the higher-order problem of the physico-chemical conditions which have to be fulfilled in order for a mutation to be capable of reflection in macroscopic change. The emphasis here is upon the word "capable", for the deterministic treatment ignores the fact that the amplification of a mutant is itself subject to statistical fluctuations. Since each mutant first appears as a single copy, fluctuations may be expected to influence its concentration until this reaches macroscopic values. The statement of deterministic theory that every better-endowed mutant I_{m+1} will *certainly* assert itself against its master sequence I_m is only true in the limit of large numbers of copies.

We must therefore address ourselves to the question of how far the results of deterministic selection theory are modified by consideration from a *stochastic** viewpoint. We shall draw in particular upon the fundamental researches of A.F. Bartholomay (1958a,b; 1959) and D.A. McQuarrie (1967), in which general methods of probability theory have been applied to problems in reaction kinetics.

---

*Stochastic theory: the application of probability theory to dynamical processes.

6.1. Fluctuations and Stability

In order to show how far limits are set upon the deterministic description of molecular systems we shall discuss two simple game models (Eigen and Winkler-Oswatitsch, 1975). The point at issue will be: "What effect do fluctuations have upon the stability of molecular systems?" It will be shown that certain stationary states - as opposed to genuine equilibria - are metastable with respect to the appearance of fluctuations.

Our game models will be based upon a 64-square board (Fig. 6.1). Each square is defined by two co-ordinates. Two appropriately inscribed octahedra can be cast so as to generate randomly the co-ordinates of any of the 64 squares. The

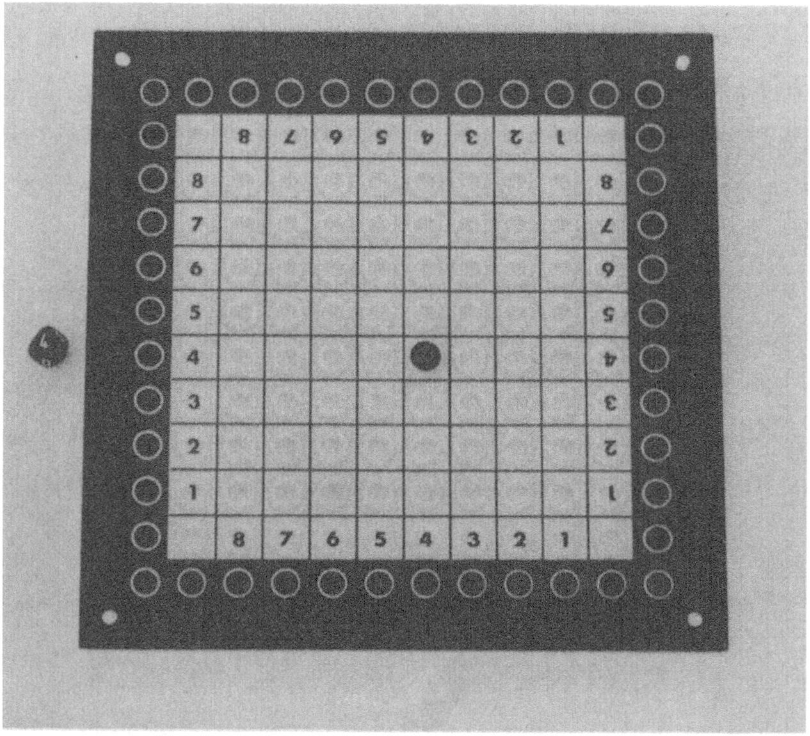

Fig. 6.1. Arrangement for the simulation of stochastic processes in a game model. The board is divided into 8 x 8 = 64 squares, each defined by co-ordinates. The squares are occupied by variously-coloured balls. Any square can be selected at random by a "throw" of two octahedral dice. The fate of a ball whose square is thrown is decided in each case by the rules of the game (see text). (Reproduced with kind permission of M. Eigen and R. Winkler-Oswatitsch.)

squares are occupied by variously-coloured balls. The fate of a ball whose co-ordinates are thrown by the dice is determined by the rules of the game. Before describing these rules we will first give a physical interpretation to the elements of this game model.

The board represents the reaction vessel of Section 4 (Fig. 4.1). The balls of various colours stand for molecular information carriers, *i.e.*, for nucleic acids with various nucleotide sequences. The dice and the rules are respectively the elements of chance and of regular behaviour in Nature. Just as in a real biochemical process, the balls will be subject to continuous processes of reproduction and disappearance. The finite size of the board will place a limit on the growth of the population and thus simulates the CP constraint (see Section 5.1). In this way we have turned our reaction vessel of Section 4 into a game-theoretical model.

(1) SELF-REGULATING FLUCTUATIONS. The first game model to be discussed goes back to the Ehrenfest urn model, well-known in physics (Ehrenfest and Ehrenfest, 1907; Kohlrausch and Schrödinger, 1926). The Ehrenfest game simulates fluctuations in systems close to stable equilibrium.

To start, the board is covered completely with, say, black and red balls, whereby their distribution on the board is arbitrary. We can start with any ratio of black to red balls. However, the principle of the Ehrenfest game becomes especially clear if one starts with only one sort of balls. We define the rules as follows: *Each ball whose co-ordinates are thrown must be removed from the board and replaced by a ball of the other colour.*

The progress of a typical Ehrenfest game, simulated on a computer, is shown in Fig. 6.2. The game always ends with the same result: independently of the initial distribution, after a certain number of throws an approximately equal distribution of black and red balls is reached. This equilibrium is stable, even though it is subject to fluctuations. As the game progresses the average rates at which black and red balls are replaced by others of the opposite colour become equal.

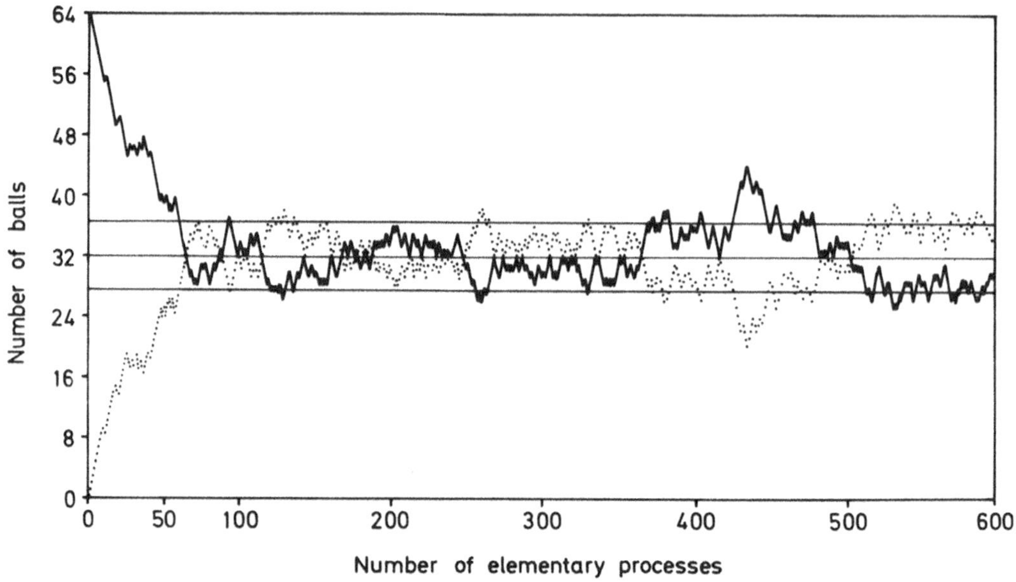

Fig. 6.2. Self-regulating fluctuations. The fluctuations about an equal distribution (Z/2) are self-regulating (half-width $\sim\sqrt{Z}$). The system approaches a state of stable equilibrium. (After Eigen, 1976.)

The result is of course predetermined by the rules. Whenever fluctuations lead to a deviation from the equilibrium distribution, the probability of reduction of the deviation grows in proportion to the size of the deviation itself. This is the phenomenon of mass-action; the Ehrenfest game does indeed simulate the equilibrium behaviour to be expected from the law of mass-action applied to a chemical reaction of the form $A \rightleftharpoons B$. The fact that we can without contradicting the above argument consider the law of mass-action in chemistry as a deterministic law becomes clear from a numerical example. A Z-value which the chemist commonly encounters is Avogadro's number ($\approx 10^{24}$). The magnitude of the half-width of fluctuations around equilibrium is about 10^{12}. However the *relative* deviation from the equilibrium distribution is only of the order of 10^{-12}. Thus deviations from the law of mass-action cannot be observed macroscopically.

Let us express the result of the equilibrium game in mathematical form. If we denote with Z the total number of red and black balls and with n the deviation at any given moment from an even distribution, then the population numbers z_1 and z_2 of red and black balls are given by

$$z_1 = \frac{Z}{2} + n \qquad \text{and } z_2 = \frac{Z}{2} - n \tag{6.1}$$

where n can take on any value from -Z/2 to +Z/2. A simple combinatorial consideration shows that the probability of a fluctuation of size n about the equilibrium value ($z_1 = z_2 = Z/2$) is

$$P(n) = 2^{-Z} \frac{Z!}{(\frac{Z}{2} + n)!\ (\frac{Z}{2} - n)!} \tag{6.2}$$

For large values of Z this probability distribution tends to the Gaussian form

$$P(n) \rightarrow \frac{\exp\left[-\left(\frac{n}{\sqrt{Z/2}}\right)^2\right]}{\sqrt{\pi(Z/2)}} \tag{6.3}$$

which is symmetric about n = 0 and has a half-width proportional to $\sqrt{Z/2}$. From Eq. (6.2), the probability of an *extreme fluctuation* in either direction ($n = \pm Z/2$) is

$$P(\frac{Z}{2}) = 2^{-Z} \tag{6.4}$$

This relation shows clearly that an extreme fluctuation is rare for large populations (large Z). We thus find confirmation of our earlier conjecture (Chapters 3 and 4) that equilibrium statistics applied to thermodynamically closed systems do not lead to the selection of any particular state, at least within physically reasonable time-scales. For example, in our game model a selected state of, say, only red balls occupying the board will on average be reached only after $2^{64} \approx 10^{19}$ throws of the dice, since each of the two possible extreme states of occupation is one of 2^{64}

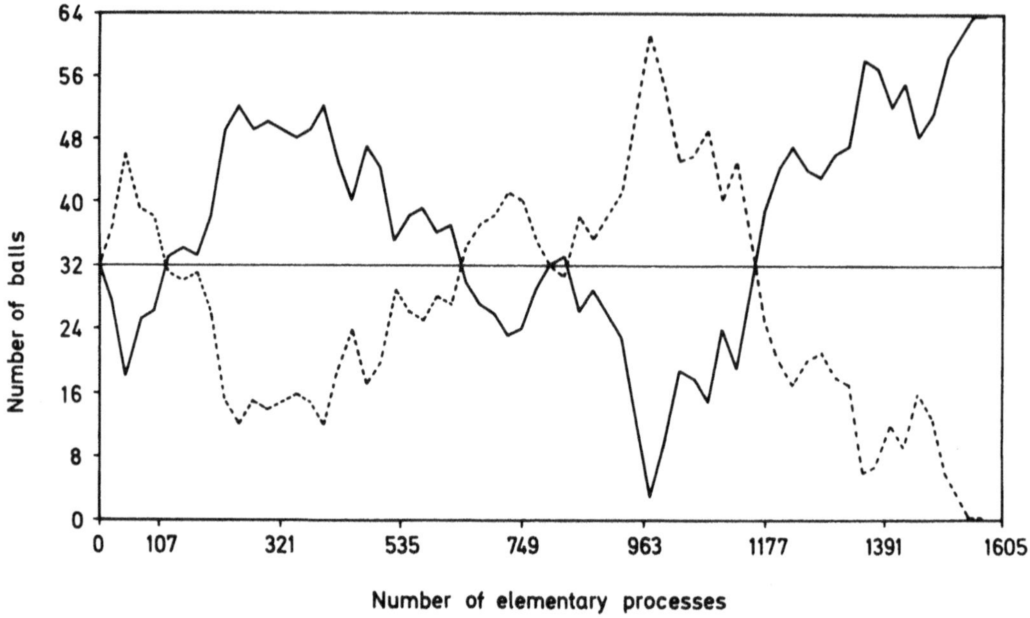

Fig. 6.3. Non-self-regulating fluctuations. The system behaves indifferently with respect to fluctuations, that is, fluctuations are neither damped nor reinforced. Since no distribution is preferred, the system diffuses through all possible population states. (After Eigen, 1976.)

possible distributions. At a rate of one throw every second, the present age of the universe ($\approx 10^{17}$s) would fall far short of ensuring that such a state is reached.

(2) NON-SELF-REGULATING FLUCTUATIONS. The equilibrium resulting from the Ehrenfest game was stable because the fluctuations were *self-regulating*: the greater the deviation, the greater the probability of reducing it. That is, the tendency was always to equalize the distribution.

We now introduce a game in which the fluctuations are *not* self-regulating. The random element here is the tossing of a coin: *If it comes down "heads", any ball of a given colour - say black - is removed from the board and replaced by a red ball. If the coin comes down "tails" the opposite action is performed.*

The probability of hitting a particular colour is now always 50%, and this is independent of the past progress of the game and of the particular momentary ditribution of the balls. Since the replacement probabilities in each direction are inherently equal, the system should (according to the deterministic view) be in a stationary state. A deterministic treatment of the system would result in a time-independent distribution, *viz.*, that which was present at the beginning of the game.

The computer simulation, however, leads to a completely different result (Fig. 6.3). The initial distribution is shown to be metastable. Since the probabilities of reduplication and of disappearance of a ball are independent of each other, the fluctuations no longer have the self-regulatory character which they had in the Ehrenfest game. As the game progresses, the distribution of balls drifts in a random walk through all possible population states. In agreement with Einstein's displacement-squared diffusion relation, on average $(Z/2)^2 = 32^2 \approx 1000$ throws are needed to proceed from an equal distribution to one of the two extreme states. The limits of a deterministic description and the necessity of a stochastic analysis are particularly clear in this example.

6.2. Markovian Processes

In the game models discussed above, the total population of balls was kept constant by assuming that in each elementary process the decrease in the population of one colour was compensated for by an increase in the population of the other. Without this CP constraint there would certainly have been a fluctuation catastrophe after a finite number of moves, *i.e.*, the population as a whole would have become extinct. This is easy to see, as long as the probabilities of birth and death are independent of each other and do not change with time. To begin with, reproduction and disappearance balance each other. With passing time the fluctuations will accumulate and at some point a negative deviation from the average value will equal the size of

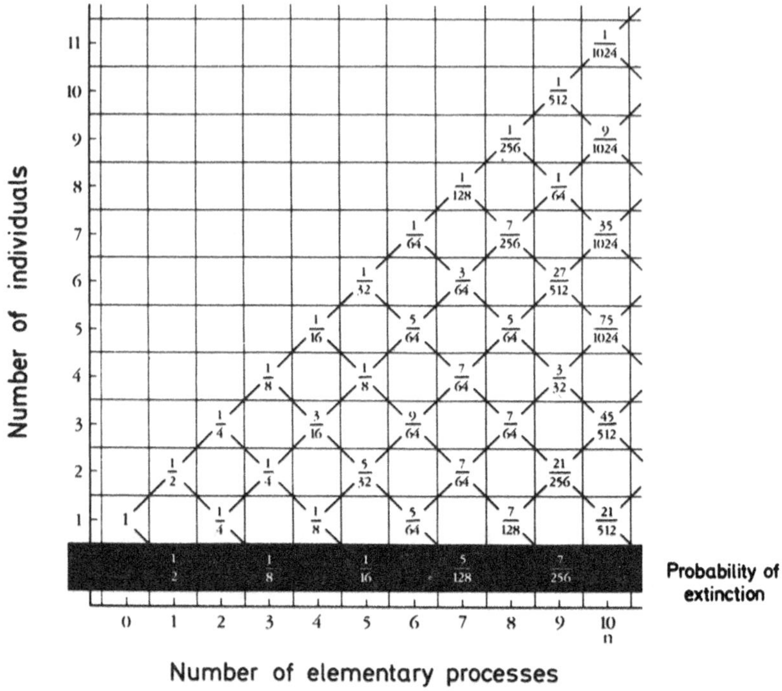

<u>Fig. 6.4.</u> Probability distribution for a linear birth-and-death process starting from one individual. Each step is a change in the population number of plus one (birth) or minus one (death), with equal probabilities. The numbers at the branching-points in the diagram give the respective probabilities of reaching these points.
It is easy to estimate after how many steps a population which at the start consists of k individuals dies out in a fluctuation catastrophe. Consider a random walk up and down the ordinate, starting from the point giving the population size k. The solution is well known: extinction (k=0) is reached on average after k^2 steps. (After Schuster, 1972.)

the population, which means extinction of the species. A typical probability distribution for this is shown in Fig. 6.4. Stochastic processes of this kind are called *linear birth-and-death processes*. They were first calculated explicitly for chemical reaction systems by A.F. Bartholomay (1958a,b).

We shall present these ideas in a precise mathematical way. But first we have to find out in what way the kinetic equations which describe the time-dependent development of a molecular system are modified by the stochastic treatment. To do this we must treat the dynamics of the system as a *Markovian process*, that is, we must attempt to find for all time t an equation which describes the further development of the system independently of its previous history. For a detailed treatment of the theory of stochastic processes see, for example, J.L. Doob (1953) and J.A. Rosanov (1975).

(1) DIFFERENTIAL EQUATIONS FOR THE PROBABILITIES OF TRANSITION. Let $\xi=\xi(t)$ be a random variable which adopts at any time t a value 0,1,2,.....,N. A stochastic process

$$\{\xi(t),\ 0 \leq t < \infty\} \tag{6.5}$$

is called a *homogeneous Markovian process*, if for any time t_1 and for an arbitrarily given value $\xi(t_1) = z$ the value of the random variable $\xi(t)$ is at any time $t \geq t_1$ independent of its value at any time $t \leq t_1$. In other words: starting from a known state $\xi(t_1) = z$ the behaviour of the system after the point in time t_1 does not depend upon its previous development up to the moment t_1.

We shall now assume that the linear birth-and-death process to be investigated can be described by a homogeneous Markovian process. The random variable $\xi=\xi(t)$ is then the population number of the system and corresponds to the discrete population states $S_0, S_1, S_2, \ldots\ldots, S_N$. The transition of the population from the state S_i to the state S_j is given by the transition probability $p_{ij}(t)$. The magnitude of $p_{ij}(t)$ thus tells us with what probability a system which at time τ is in state S_i will between times τ and $\tau+t$ have made the transition to state S_j. The transition probability is independent of τ and is defined as the conditional probability

$$p_{ij}(t) = P\{\xi(t+\tau)=j \mid \xi(\tau)=i\} \qquad (i,j = 0,1,2,\ldots) \tag{6.6}$$

that the random variable ξ has the value j if it had the value i a time t ago.

The transition probabilities $p_{ij}(t)$ possess certain important properties (*cf.* Rosanov, 1975). For example, it is apparent that

$$p_{ik}(t+\tau) = \sum_j p_{ij}(\tau)p_{jk}(t) \qquad \text{(for arbitrary } t,\tau \geq 0) \tag{6.7}$$

$$p_{ij}(0) = \begin{cases} 1 \text{ for } i=j \\ 0 \text{ for } i \neq j \end{cases} \tag{6.8}$$

Let us now consider the time behaviour of the transition probabilities. The probability that *no* transition takes place in a small time interval Δt is given by $p_{ii}(\Delta t)$, so that $1 - p_{ii}(\Delta t)$ is the probability for a "jump" from state S_i into one of the other states $S_{j\neq i}$. This transition probability is assumed to be proportional to Δt, *i.e.*,

$$1 - p_{ii}(\Delta t) = q_i\Delta t + o(\Delta t) \tag{6.9}$$

where $o(\Delta t)$ contains the higher-order Δt terms, *i.e.*, $\lim_{\Delta t\to 0} o(\Delta t)/\Delta t = 0$. Since the probability of transition from S_i to $S_{j\neq i}$ is proportional to Δt we have

$$p_{ij}(\Delta t) = q_{ij}\Delta t + o(\Delta t) \tag{6.10}$$

If we define q_{ii} as $-q_i$ then we can construct a matrix $\mathbf{Q}$ with constant elements taken from Eqs. (6.9) and (6.10):

$$\mathbf{Q} = (q_{ij}) = \begin{cases} \lim\limits_{\Delta t\to 0} \dfrac{p_{ii}(\Delta t)-1}{\Delta t} = \dot{p}_{ii}(0) & \text{for } i=j \\ \lim\limits_{\Delta t\to 0} \dfrac{p_{ij}(\Delta t)}{\Delta t} = \dot{p}_{ij}(0) & \text{for } i\neq j \end{cases} \tag{6.11}$$

These conditions and assumptions lead to the following set of differential equations for the probability function (for their derivation see Appendix A.4):

$$\dot{p}_{ik}(t) = q_{kk}p_{ik}(t) + \sum_{j\neq k} q_{jk}p_{ij}(t) \qquad \text{(forward system)} \tag{6.12}$$

$$\dot{p}_{ik}(t) = q_{ii}p_{ik}(t) + \sum_{j \neq i} q_{ij}p_{jk}(t) \qquad \text{(backward system)} \qquad (6.13)$$

Equations (6.12) and (6.13) are called the *Kolmogorovian forward* and *backward equations*. The forward equation describes what has happened in the last interval of time before the transition, while the backward equation describes the development of the system in the first interval of time after the transition. This is expressed by the fact that in the forward system the final state S_k is varied, whereas in the backward system the initial state S_i is varied (*cf.* Definition 6.11). We now apply this result to the problem of linear birth-and-death processes.

(2) LINEAR BIRTH-AND-DEATH PROCESSES. In the deterministic theory, such processes can be described by a differential equation of the type

$$\dot{z} = Fz - Dz = (F-D)z \qquad (6.14)$$

where F and D are constant parameters of growth and decay and z is the number of individuals at time t.

If the initial conditions $z(0) = z(t=0)$ and the constants F and D are given, the differential equation (6.14) enables us to calculate the value of the population size z at every value of the time t. Integration of Eq. (6.14) leads to the deterministic solution:

$$z(t) = z(0)\, e^{(F-D)t} \qquad (6.15)$$

In reality, the population size is of course given by an integer, so that to represent z(t) as a continuous function is only an approximation. Further, one should strictly speaking regard z(t) as a random variable, since - as already mentioned - it is subject to irregular fluctuations.

We shall therefore construct for the process described in Eq. (6.14) a stochastic model capable of describing the fluctuations around the deterministic solution (6.15).

Assuming that the above process represents a homogeneous Markovian process, the appropriate $\mathbf{Q}$-matrix (Definition 6.11) can be constructed in the following way:

According to the deterministic solution (6.15), the relative change in the variable $z = z(t)$ in a small interval Δt is given by

$$\frac{z(t+\Delta t)-z(t)}{z(t)} = \frac{\Delta z}{z} = \exp\left[(F-D)\Delta t\right] - 1 \tag{6.16}$$

Expanding the function $\exp\left[(F-D)\Delta t\right]$ in a power series,

$$\exp\left[(F-D)\Delta t\right] = \sum_{n=0}^{\infty} \frac{\left[(F-D)\Delta t\right]^n}{n!} = 1 + (F-D)\Delta t + \frac{(F-D)^2(\Delta t)^2}{2!} + \ldots \tag{6.17}$$

we obtain (for small values of Δt)

$$\frac{\Delta z}{z} = F\Delta t - D\Delta t + o(\Delta t) \tag{6.18}$$

where $o(\Delta t)$ again denotes the sum of all the remaining terms of higher order. This term vanishes for $\Delta t \to 0$. If we now denote with Δf the number of reduplication events and with Δd the number of disappearances in the time interval Δt, Eq. (6.18) can be written in the form

$$\frac{\Delta z}{z} = \frac{\Delta f}{z} - \frac{\Delta d}{z} \tag{6.19}$$

where the first term on the right-hand side gives the probability with which an individual duplicates itself in the interval Δt and the second term represents the probability of disappearance of an individual.

The probability that a population of z individuals increases or decreases by exactly one in the time interval Δt can now easily be calculated. We set out from the fact that only one of three things can happen to an individual during the interval Δt: it can (i) double itself, or (ii) be destroyed, or (iii) neither double itself nor be destroyed. The probability of exactly one birth event in the interval Δt can then be shown to be

$$p_{z,z+1} = Fz\Delta t + o(\Delta t) \tag{6.20}$$

and the probability of exactly one death event in the same time is

$$p_{z,z-1} = Dz\Delta t + o(\Delta t) \tag{6.21}$$

These relations follow from the binomial distribution

$$B(k,m,p) = \binom{m}{k} p^k (1-p)^{m-k} \tag{6.22}$$

for $k = 1$ successes in $m = z$ trials with the probabilities $p_f = \Delta f/z$ or $p_d = \Delta d/z$ (for the derivation see Appendix A.4).

We choose the time interval to be small enough to describe completely the changes in the population size by the probabilities $p_{z,z+1}$ and $p_{z,z-1}$. Therefore in the time interval Δt (independently of the starting-point t) only the transitions $S_i \rightarrow S_{i-1}$ or $S_i \rightarrow S_{i+1}$ are possible ($i = 1,2,\ldots$), where i is the size of the population at the beginning of the interval Δt. Each transition takes place in a single elementary step, that is, the possibility of a succession of events within the interval Δt, leading to a net change in the population of exactly one individual, will be discounted. In addition, the transition probability for the transition $S_0 \rightarrow S_1$ is zero - in other words, the system dies out as soon as it reaches state S_0.

It is now easily shown that the relations (6.20) and (6.21) suffice to determine the components of the $\mathbf{\Omega}$-matrix. For this purpose we expand the probability function in a Taylor series about the point t=0, so that

$$p_{ij}(\Delta t) = p_{ij}(0) + \dot{p}_{ij}(0)\Delta t + o(\Delta t) \tag{6.23}$$

From this it follows, taking into consideration the initial conditions (6.8) and Definition (6.11) that for $j = i+1$

$$p_{i,i+1}(\Delta t) = q_{i,i+1}\Delta t + o(\Delta t) \tag{6.24}$$

and for $j = i-1$

$$p_{i,i-1}(\Delta t) = q_{i,i-1}\Delta t + o(\Delta t) \tag{6.25}$$

Comparing Eqs. (6.24) and (6.25) with Eqs. (6.20) and (6.21) the components of the $\mathbf{Q}$-matrix are obtained

$$q_{i,i+1} = iF \qquad (i = 0,1,2,.....) \tag{6.26}$$

$$q_{i,i-1} = iD \qquad (i = 1,2,3,.....) \tag{6.27}$$

Finally, since

$$p_{i,i} = 1 - p_{i,i+1} - p_{i,i-1} \tag{6.28}$$

it follows that

$$q_{i,i} = -i(F+D) \qquad (i = 0,1,2,.....) \tag{6.29}$$

Inserting the appropriate elements of the $\mathbf{Q}$-matrix into the Kolmogorovian equations (6.12) and (6.13) we obtain the following system of differential equations for the transition probabilities of the linear birth-and-death-process:

$$\dot{p}_{ik}(t) = -k(F+D)p_{ik}(t) + (k-1)Fp_{i,k-1}(t) + (k+1)Dp_{i,k+1}(t) \tag{6.30}$$

(forward system)

and

$$\dot{p}_{ik}(t) = -i(F+D)p_{ik}(t) + iFp_{i+1,k}(t) + iDp_{i-1,k}(t) \tag{6.31}$$

(backward system)

This is a system of ordinary differential equations which can be transformed into a partial differential equation with the help of a generating function. The detailed procedure for the solution is described in Appendix A.4. Two cases should be distinguished:

(a) *Equal growth and decay parameters* (F=D). If the parameters of growth and decay are equal, the solution obtained for the transition probabilities $p_{ik}(t)$ is

$$p_{ik}(t) = \sum_{n=0}^{k \text{ or } i} (-1)^n \binom{i}{n} \binom{i+k-n-1}{k-n} (Ft)^{i+k-2n} (Ft-1)^n (Ft+1)^{-i-k+n} \tag{6.32}$$

where the upper limit of summation is k if $0 < k < i$, and i if $k \geq i$ (see Appendix A.4).

(b) *Differing growth and decay parameters* (F≠D). If the parameters of growth and decay are different, the following solution is obtained for the transition probabilities $p_{ik}(t)$:

$$p_{ik}(t) = \sum_{n=0}^{k \text{ or } i} (-1)^n \binom{i}{n} \binom{i+k-n-1}{k-n} F^{k-n} D^{i-n} \cdot \left[e^{(F-D)t}-1\right]^{i+k-2n} \left[Fe^{(F-D)t}-D\right]^{-i-k+n} \left[De^{(F-D)t}-F\right]^{n} \tag{6.33}$$

where the upper limit of summation is k if $0<k<i$ and i if $k \geq i$ (see Appendix A.4).

The reader can easily ascertain that in both cases the boundary conditions (6.8) are fulfilled for the function $p_{ik}(t)$.

6.3. Stochastic Treatment of Simple Selection Models

Having demonstrated by arguments from simple game models that a stochastic analysis of the selection problem is necessary, we shall now study two concrete selection models, due to M. Eigen (1971) and P. Schuster (1972), and use these to evaluate the results of the foregoing section. To do this we return to the model system of Chapter 4 and Fig. 4.1.

Let there be a large number Z of different molecular species I_i. (*N.B.* The subscripts i and j now refer as usual to species and not to population states.) Let each species at the start be represented by one copy only. In accordance with the assumptions of our model system (*cf.* Eq. 4.5), the formation of a species is determined solely by a template-instructed copying reaction. Without any further constraints the phenomenological rate equations are then (see Eq. 5.10):

$$\dot{x}_i = (F_i - D_i^{\varphi})x_i + \sum_{j \neq i} \psi_{ij} x_j \qquad (i = 1, \ldots\ldots, N) \tag{6.34}$$

where use has been made of the abbreviation

$$D_i^{\varphi} = D_i + \varphi \tag{6.35}$$

We shall discuss two cases involving different assumptions about the kinetic parameters:

MODEL 1. All parameters F_i and D_i^{φ} are constant and equal:

$$F_i \equiv F = D_i^{\varphi} \equiv D^{\varphi} = \text{const.} \tag{6.36}$$

Further, let the reproduction of all the molecular species I_i be error-free, so that

$$Q_i = 1 \qquad \text{and} \qquad \psi_{ij} = 0 \qquad (\text{for all } i,j) \tag{6.37}$$

Our model thus corresponds to the linear birth-and-death process described in Section 6.2, with $F \equiv D$. The deterministic rate equation (6.34) has in this case the simple form

$$\dot{x}_i = (F - D^{\varphi})x_i \equiv 0 \tag{6.38}$$

That is, the concentrations of all the species are constant:

$$x_i(t) = x_i(0) = \text{const.} \tag{6.39}$$

Thus the macromolecular composition of our model system is, according to the deterministic theory, also invariant in time.

A stochastic approach, on the other hand, leads to a completely different result. Starting from a uniform class of Z individuals at time t=0, the probability of extinction of the population follows directly from Eq. (6.32):

$$p_{Z0}(t) = \left(\frac{Ft}{1+Ft}\right)^Z \tag{6.40}$$

This relationship confirms our previous reasoning and shows clearly the limitation of deterministic theory. In particular, in the limit of large t, we have

$$\lim_{t\to\infty} p_{Z0}(t) = 1 \tag{6.41}$$

That is, our system is certain to die out at some time, irrespective of the magnitude of Z at time t=0.

We can also predict the time within which the fluctuation catastrophe may be expected to occur. At time $t = Z/F$ an approximate solution for large Z is

$$p_{Z0}(t=Z/F) = \left(\frac{Z}{1+Z}\right)^Z \to \frac{1}{e} \tag{6.42}$$

That is, the probability that extinction will occur sometime during this period is about 37%*. At a time $t = Z^2/F$ we obtain

$$p_{Z0}(t=Z^2/F) = \left(\frac{Z^2}{1+Z^2}\right)^Z \to 1 - \frac{1}{Z} \tag{6.43}$$

so that by this time extinction is almost certain.

We can therefore conclude that in contrast to the deterministic result (6.39), a population consisting of Z individuals at time t=0 will die out on average at some time between $t = Z/F$ and $t = Z^2/F$ (see also the discussion of Fig. 6.4).

For large populations the stochastic and the deterministic theories should lead to the same result. The time required for a *large* population to die out has been estimated by P. Schuster (1972), assuming realistic values for F. For example, we know from experiment that the rate constants of unimolecular reactions rarely exceed a value of $10^8\ s^{-1}$. Thus for chemically relevant population sizes, for example with $Z = 6.02 \cdot 10^{23}$ (Avogadro's number), the population would die out within $t_1 = Z/F \approx 10^{15}$ s to $t_2 = Z^2/F \approx 10^{39}$ s. This example confirms our expectation that for large numbers of individuals the stochastic modification of deterministic kinetics makes no appreciable difference.

Our analysis has so far been restricted to the fate of the entire population represented by the number Z. In order to establish the stochastic distribution of individuals among different molecular species we shall now examine the fate of the individuals. We first illustrate this with a simple numerical example. Consider a population of $Z = 10^6$ individuals, subdivided into 10^3 classes of $z = 10^3$ members each. All reproduce and decay independently with equal probability.

After a time $t = Z/F = 10^6/F$ the population as a whole has a probability of survival of $1-p_{10^6,0} = 0.63$ (see Eq. 6.42). At the same time $(t = z^2/F = 10^6/F)$, however, the probability of survival for each class separately is only 0.001 (see

---

*A more rigorous treatment of the stochastic problem yields the extinction time $t = (Z-1)/F$, which of course is nearly identical with $t = Z/F$ for large Z values (Schuster and Sigmund, 1982).

Eq. 6.43). This means that on average the individuals remaining will all belong to the same class.

From Eq. (6.32), the probability that a particular individual is represented k times after a time t is given by

$$p_{1k}(t) = \frac{1}{(Ft)^2}\,\frac{1}{(1+1/Ft)^{k+1}} \qquad (k \geq 1) \tag{6.44}$$

and the probability that it dies is

$$p_{10}(t) = \frac{Ft}{1+Ft} \tag{6.45}$$

After many trials, each starting from one individual, the average distribution of population sizes is given by the function $kp_{1k}(t)$. Therefore a large number Z of distinct individuals has a distribution of class sizes at time t given by $Zkp_{1k}(t)$. (Here "class" implies having a common ancestor.) The function kp_{1k} has a maximum with respect to k at

$$k_{max} = \frac{1}{\ln(1+1/Ft)} \tag{6.46}$$

For $Ft \gg 1$ we use the approximation $\ln(1 + 1/Ft) \approx 1/Ft$ and obtain

$$k_{max} \approx Ft \tag{6.47}$$

and

$$(kp_{1k})_{max} = \frac{1}{Ft} \exp\left[-(1 + \frac{1}{Ft})\right] \approx \frac{1}{Ft} \cdot \frac{1}{e} \tag{6.48}$$

The distribution function $kp_{1k}(t)$ is plotted in Fig. 6.5 as a function of k at times $t = 0.2/F$, $t = 1/F$ and $t = 5/F$. The trend towards larger sizes with increasing time is apparent.

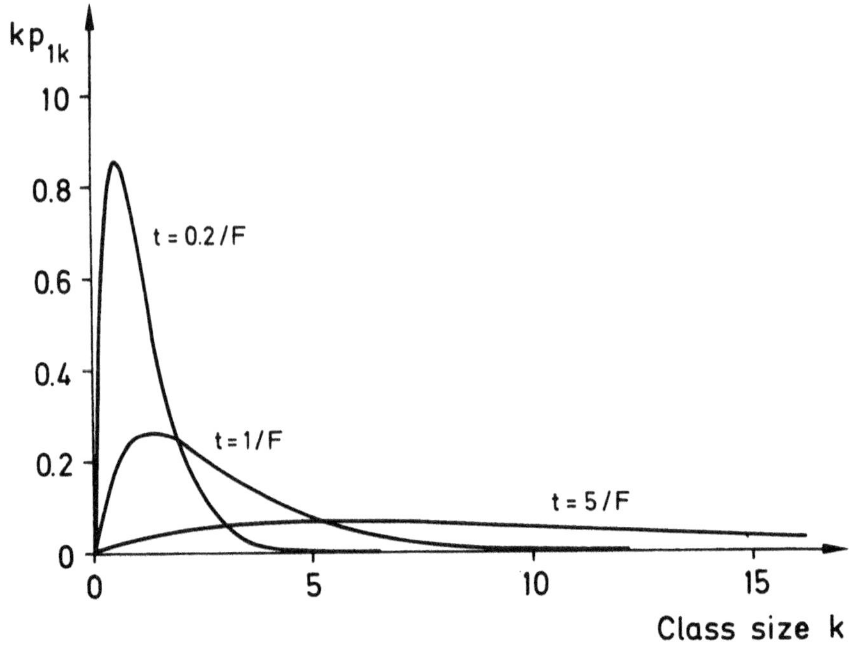

Fig. 6.5. Distribution of sizes of the polymer classes surviving at times t = 0.2/F, t = 1/F and t = 5/F. The distribution is given by $kp_{1k}(t)$ (*cf.* Eqs. 6.44 - 6.48) and has a maximum at approximately k = Ft; that is, individuals present are most likely to belong to a class with z = Ft members. For $t \geq Z/F$ we have $k = Ft \geq Z$, so that the survivors are very probably all identical.

Let us summarize our results up to now. The system always loses information when an individual information carrier is lost before it can reproduce itself. The number of distinct, *i.e.*, non-redundant information carriers therefore decreases continually, tending to reduce the overall population in our model system. At the same time the redundancy of the species surviving rises continually. Finally the entire population dies out. For t = Z/F the maximal probability is at k=Z (see Fig. 6.5). However, at this point the population still has a high probability of survival (see Eq. 6.42), so it is very likely to be composed uniformly of one species. The tautology of the "survival of the survivors" is expressed clearly in this model: the attribute of "fitness" is determined solely by the fact of having survived, since each individual has the same chance *a priori* of being present, with maximal redundancy, at the time t = Z/F.

MODEL 2. We shall now analyse a model in which the parameters of growth and decay are constant but different from each other:

$$F_i \neq D_i^{\varphi} \tag{6.49}$$

For simplicity's sake we again assume that the reproduction of every molecular species I_i is error-free:

$$Q_i = 1 \qquad \text{and} \qquad \psi_{ij} = 0 \qquad \text{(for all i,j)} \tag{6.50}$$

The deterministic rate equations are then

$$\dot{x}_i = (F_i - D_i^{\varphi})x_i \qquad (i = 1,\ldots\ldots,N) \tag{6.51}$$

with the solutions

$$x_i = x_i(0)\,\exp\left[(F_i - D_i^{\varphi})t\right] \qquad (i = 1,\ldots\ldots,N) \tag{6.52}$$

Without population control the system can only remain in a stationary state as long as the overall decay and growth rates are mutually compensating. The indefinite maintenance of a stationary state requires mechanisms for control and regulation such as those discussed in Chapter 5. However, at the moment we are more interested in the question of what happens to a mutant which is better-equipped but which is present initially as only one copy.

The deterministic equation (6.51) predicts that each species for which the rate of formation exceeds the rate of decay increases in concentration. We shall now compare this with the stochastic prediction.

What is the probability that a species I_i, represented at time t=0 by k copies, will die out in consequence of a fluctuation catastrophe? For differing growth and decay parameters ($F_i \neq D_i^{\varphi}$) this probability is obtained from Eq. (6.33):

$$p_{k0}(t) = \left(1 - \frac{F_i - D_i^{\varphi}}{F_i - D_i^{\varphi}\exp(D_i^{\varphi} - F_i)t}\right)^k \tag{6.53}$$

or, in the limit of large t,

$$\lim_{t\to\infty} p_{k0} = \left(\frac{D_i^{\varphi}}{F_i}\right)^k \tag{6.54}$$

Thus the chance of survival is, in contrast to the deterministic approach, not simply unity, but only

$$1-p_{k0} = 1 - \left(\frac{D_i^{\varphi}}{F_i}\right)^k < 1 \qquad (\text{for } F_i > D_i^{\varphi}) \tag{6.55}$$

That is, even for $F_i > D_i^{\varphi}$ the species only has a limited chance of survival.

What conclusions can be drawn about selection kinetics? Assume that in selection equilibrium a mutant I_{m+1} with a selective advantage $W_{m+1} > W_m$ appears. Since each mutant is at first present in one copy only, its probability of survival is at the moment of its appearance

$$1-p_{10} = 1 - \frac{D_{m+1}^{\varphi}}{F_{m+1}} \tag{6.56}$$

In accordance with Eq. (6.55), the probability of survival of the mutant I_{m+1} grows as the number of its progeny increases. At high k values the probability of extinction $p_{k0}(t)$ approaches asymptotically the value of zero.

The stochastic model therefore shows that there is no point in time after which a selectively favourable mutant is bound to assert itself. Even for high population numbers k, there is always a finite (if small) probability that a superior mutant will die out in a fluctuation catastrophe.

Table 6.1 shows (i) the probabilities of survival of mutants which according to deterministic theory would be certain to survive and (ii) the number of copies which each mutant would have to reach in order to have a survival probability of at least (1 - 1/e) = 63%.

Ratio of rate constants F_i/D_i^{φ}	Probability of survival for a single copy	Number of copies which together have a probability of survival of at least (1-1/e)
1.0001	10^{-4}	10^4
1.001	10^{-3}	10^3
1.01	10^{-2}	10^2
1.1	0.09	11
1.3	0.23	4
1.5	0.33	3
2	0.5	1
5	0.8	1
10	0.9	1

Table 6.1. Stochastic thresholds for the survival of mutants. A mutant which has only a small selective advantage over a dominant species has scarcely any chance of survival. In this way a stochastic analysis leads to a result qualitatively different from that of the deterministic approach. (After Schuster, 1972.)

These conclusions are based upon the kinetic equations (6.51), and consequently apply only to the linear case. The stochastic theory of a non-linear selection model has been developed by P.H. Richter (1975) and by B.L. Jones and H.K. Leung (1981). As a mathematical treatment of non-linear models is relatively cumbersome and requires many simplifying assumptions, no attempt will be made to present it here.

III. The Transition from the Non-Living to the Living

7. The Information Threshold

An examination of the nucleic acids present in living systems reveals a clear connection between the complexity of a living being and its capacity for information storage: the more complex a living system, the greater the symbol content of its information carrier. This fact presents a problem in connection with the origin of genetic information. On the one hand, evolutionary progress is most rapid if the mutation rate within a population of information carriers, and thus the genetic variability, is high. On the other hand, a high mutation rate must lead to a large number of "mistakes" in each nucleic acid chain copied and thus to a loss of information.

The problem can be re-formulated as follows. An organism may well increase its chances of survival by becoming more complex. This calls for a larger information store, which however will be more prone to errors in copying.

The question therefore arises: What is the greatest chain length of a nucleic acid molecule which can be reproducibly transmitted under selection pressure over an indefinite number of generations? This fundamental question will now be formulated more precisely and investigated in detail. We shall see that a new principle of organization is necessary to take evolution beyond a certain point.

7.1. The Quality Function

Let us consider the replication of a nucleic acid molecule. During this process each base must be "recognised" by its complementary partner. We have already seen (Section 2.1) that the molecular recognition process takes place by way of a specific interaction involving hydrogen bonds. Because of the finite energy of interaction and the randomizing effect of thermal motion an absolutely error-free recognition between complementary nucleotides is impossible. In addition, the accuracy of transmission of a single symbol (nucleotide) is in general influenced by its nearest neighbours and possibly also by remoter nucleotides. We shall examine this more closely in Section 7.4.

Let us first look at the connection between the quality q of copying of single nucleotides and the quality function Q for the copying of an entire nucleotide sequence (Section 5.1). For simplicity we assume that the probability of correct incorporation of every nucleotide is independent of the nature of the neighbouring nucleotides. We shall denote the accuracy of copying of nucleotides of type μ as q_μ. The overall probability Q of error-free reproduction of an information carrier I of chain length ν is then given by the product of all the independent individual probabilities q_μ, that is

$$Q = q_A^{\nu_A} \; q_U^{\nu_U} \; q_G^{\nu_G} \; q_C^{\nu_C} \tag{7.1}$$

where ν_μ is the number of times with which the base μ appears in the information carrier concerned. The ν_μ values are related by

$$\nu = \sum_\mu \nu_\mu \tag{7.2}$$

We can define an average q value as

$$\bar{q} = \sqrt[\nu]{Q} \tag{7.3}$$

The $\bar{q}$, while not depending explicitly upon the identity of the particular nucleotide to be copied, still reflects the frequencies with which the four nucleotide types appear in the information carrier I_i. From now on we shall assume for simplicity that all the I_i have the same chain length and composition, so that $\bar{q}$ requires no index. From Eq. (7.3) we have

$$Q = Q(\bar{q},\nu) = \bar{q}^{\nu} \tag{7.4}$$

We shall restrict the following discussion to systems for which Eq. (7.4) represents a good approximation. The quality function Q depends in such cases no longer upon the individual sequence of single information carriers but only upon the chain length ν - *i.e.*, upon the information storage capacity.

If one chain is copied once then the probability of error-free copying is equal to Q. Generalizing, we now define $Q_{\nu k}$ as the probability of appearance of k errors during one copying of a chain of length ν symbols. This probability is given by the binomial distribution

$$Q_{\nu k} = Q(\bar{q},\nu,k) = \binom{\nu}{k}\bar{q}^{\nu-k}(1-\bar{q})^{k} \tag{7.5}$$

where $Q_{\nu 0} = Q$ (*cf.* Eq. 7.4). For small error probabilities $(k(1-\bar{q}) \ll 1)$ this distribution approximates to the Poisson distribution

$$Q_{\nu k} = \frac{\varepsilon^{k} e^{-\varepsilon}}{k!} \tag{7.6}$$

where the quantity

$$\varepsilon = \nu(1-\bar{q}) \tag{7.7}$$

is the *expectation value* for the appearance of exactly one error in a sequence of ν symbols.

Finally, the probability of the appearance of a *particular* mutant with k errors in k *defined* positions is

$$\tilde{Q}_{\nu k} = \frac{Q_{\nu k}}{\binom{\nu}{k}(\mu-1)^k} = \frac{\bar{q}^{(\nu-k)}(1-\bar{q})^k}{(\mu-1)^k} \to \left(\frac{\bar{q}^{-1}-1}{\mu-1}\right)^k e^{-\varepsilon} \qquad (7.8)$$

where μ is again the number of different classes of symbols and the right-hand expression is the Poisson limit for small error probabilities.

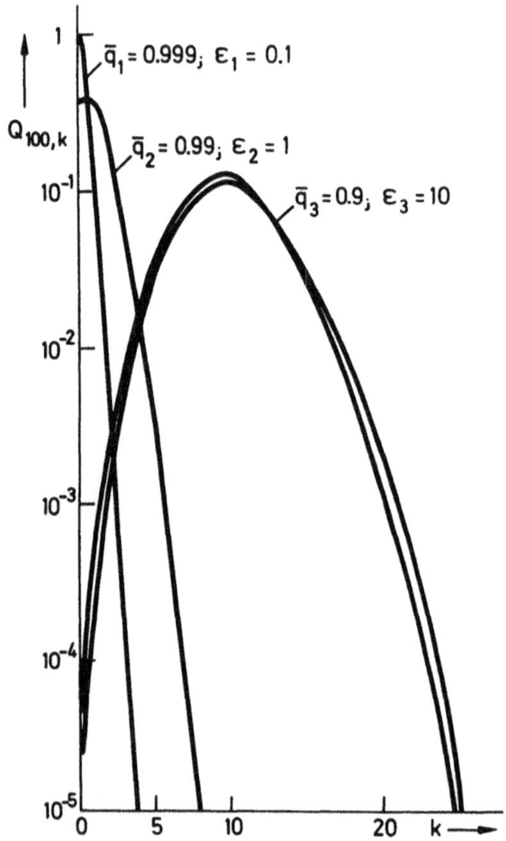

Fig. 7.1. Probability distribution for the appearance of mutants from a single copy. The quality function $Q_{\nu k}$ (Eq. 7.5) is shown for various $\bar{q}$ values with the parameter $\nu = 100$. For comparison, the Poisson distribution for $\bar{q}_3$ is shown as well.
For high $\bar{q}$ values (small error probabilities) the binomial and Poisson distributions are virtually identical for small k (the difference is too small to see). Larger deviation is seen for larger k (not shown) and for $\bar{q} = 0.9$, since then the pre-condition for a Poisson approximation $(k(1-\bar{q}) \ll 1)$ is no longer met. Each curve has a maximum at $k = \varepsilon$. (After Eigen, 1971.)

The probability distributions (7.5) and (7.6) are shown in Fig. 7.1. They express clearly the properties of the quality function Q: the smaller the expectation value ε for the appearance of errors, the greater is the accuracy of reproduction. This is manifested by a narrower $Q_{\nu k}$ distribution and a higher $Q_{\nu 0}$ value.

7.2. Conditions for Stable Selection

We shall now examine the question of how large the error rate is allowed to be before a state of organization, once attained, is lost by continual error-prone reproduction. For this purpose we consider the concentration ratio in selection equilibrium between a master sequence I_m and the stationary distribution of mutants issuing from it.

The average excess productivity $\overline{E}$ can formally be divided into two terms, $\overline{E}_m$ and $\overline{E}_{k \neq m}$, corresponding respectively to the contributions from the master sequence I_m and from all its mutants $I_{k \neq m}$. Since the reversion of a mutant into the master sequence is much less likely than the loss of information by incorrect reproduction of the master sequence, we can neglect reverse mutation. The relative stationary-state concentration of the master copy is under CP conditions:

$$\frac{x_m^s}{c} = \frac{W_m^{CP} - \overline{E}_{k \neq m}}{E_m - \overline{E}_{k \neq m}} \tag{7.9}$$

This result has been derived above (*cf.* Eq. 5.34). Rearrangement of Eq. (7.9) and use of Definitions (5.15) and (5.17) give

$$1 - \frac{x_m^s}{c} = \frac{A_m}{E_m - \overline{E}_{k \neq m}} (1 - Q_m) \tag{7.10}$$

That is, the relative concentration of mutants $(1 - x_m^s/c)$ in the stationary state is directly proportional to the error rate $(1 - Q_m)$.
From Eq. (7.10) it follows that, for $Q_m = 1$,

$$x_m^s = c \tag{7.11}$$

This is the special case of a selection process which results ultimately in the survival of only one species, *viz.*, the one with the highest selection value in the original population. In order for further evolution to take place in the system, the quality function Q must be less than unity. The greater the deviation of Q from unity, the broader will be the stationary distribution of mutants, and the greater will be the probability that among these a selectively advantageous mutant emerges (*cf.* Section 5.2). On the other hand, the inaccuracy must not be so large that the state of information represented by the master sequence "evaporates" in the course of further cycles of reproduction.

From the condition $x_m^s > 0$, Eq. (7.9) and the definition (5.17) of W_m^{CP}, the following threshold relation emerges:

$$Q_m > Q_{min} = \frac{\bar{E}_{k \neq m} + D_m}{A_m} \tag{7.12}$$

Optimal conditions for evolution clearly obtain immediately above this threshold value, where the master sequence can just hold ground against its numerous, less well-equipped mutants, but where the greatest possible number of alternative sequences can be tested.

The quantity

$$\Theta_m = Q_{min}^{-1} = \frac{A_m}{\bar{E}_{k \neq m} + D_m} \tag{7.13}$$

denotes the advantage in growth which the master sequence possesses *vis-à-vis* its mutants $I_{k \neq m}$. It is therefore reasonable to refer to Θ_m as the *superiority parameter* of species I_m.

The criterion for *stable* selection is thus:

$$Q_m > \Theta_m^{-1} \tag{7.14}$$

and this applies for every possible kind of reproductive mechanism, even if Q_{min} cannot be expressed - as in the case of quasi-linear selection equations - in the simple form of Eq. (7.12).

Combining the threshold relation (7.14) and expression (7.4) for the quality function Q, we obtain an expression for the greatest amount of information which with a given error rate can be transmitted reproducibly from one generation to the next:

$$\nu_{max} = -\frac{\ln \Theta_m}{\ln \bar{q}} \tag{7.15}$$

If $(1-\bar{q}) \ll 1$, then $\ln\bar{q} \approx \bar{q}-1$, so that

$$\nu_{max} = \frac{\ln \Theta_m}{1-\bar{q}} \tag{7.16}$$

The physical content of the threshold relationship (7.16) can be expressed in another way. If an error catastrophe is to be avoided then the expection value for the appearance of an error in a sequence of ν_{max} symbols,

$$\varepsilon_{max} = \nu_{max}(1-\bar{q}) \tag{7.17}$$

must lie below an exactly defined threshold value Θ_m:

$$e^{\varepsilon_{max}} < \Theta_m \tag{7.18}$$

This relationship follows directly from Eq. (7.16).

7.3. The Accuracy of Copying of Individual Symbols

The threshold relation (7.16) is of fundamental importance for the origin of life. We shall therefore discuss in more detail the parameters which it contains, and shall try to estimate their values on the basis of experimental data.

We consider first the accuracy parameter for single symbols, q, which we shall express by means of the free energy of base-pair interaction. To do this we consider

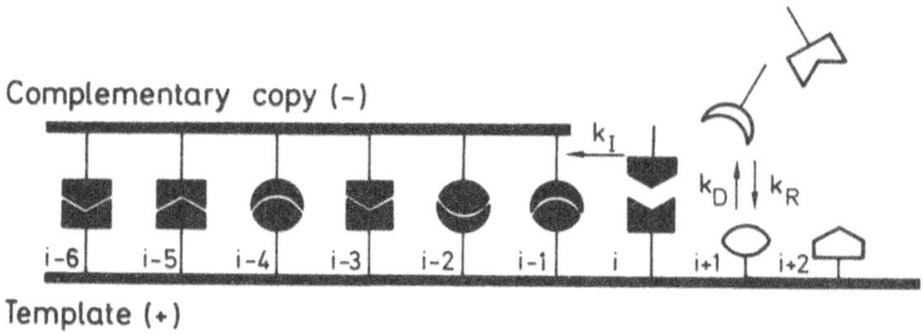

Fig. 7.2. The template-instructed synthesis of a nucleic acid takes place in two basic steps: (a) binding of the incoming nucleotide and (b) incorporation of the nucleotide into the growing nucleotide chain.
Let the rate constant for base-pair dissociation be k_D and for base-pair association k_R, and the rate constant for the covalent attachment of a nucleotide be k_I. The recognition process is described as "equilibrated" when the rate constants obey the condition $k_I \ll k_D$.

the scheme, shown in Fig. 7.2, describing the incorporation of a nucleotide into a growing nucleotide chain. If the rate-limiting step in this process is the formation of the phosphodiester bond, and not the formation of the base pair, then the recognition step preceding bond formation is in equilibrium and the accuracy parameter q can be expressed directly as a function of the free monomer concentration m_N and the stability constants of pair formation $s_{N_+N_-}$, where N_+ and N_- refer respectively to the bases of the (+) and the (-) strands. Let us examine for example the specific case where the nucleotide A is to be incorporated. If the recognition step is in equilibrium, it follows from elementary thermodynamics that

$$q = \frac{m_A s_{AU}}{\sum\limits_{N=A,U,G,C} m_N s_{NU}} \tag{7.19}$$

In general we may expect the parameters q_+ and q_- for copying the (+) and (-) strands to be the same. Should any differences arise, we can represent the accuracy parameters by their geometric mean:

$$q = \sqrt{q_+ q_-} \tag{7.20}$$

If the recognition process is not at equilibrium then the stability constants may be replaced by appropriate constants for the stationary state.

The stability constants $s_{\alpha\beta}$ can be expressed, using well-known thermodynamic relationships, in terms of the standard free energy $\Delta G_{\alpha\beta}$ of base-pairing between N_α and N_β:

$$s_{\alpha\beta} = \exp\left(-\frac{\Delta G_{\alpha\beta}}{RT}\right) \tag{7.21}$$

If the monomers are all buffered and have the same concentration then they cancel out of the fraction in Eq. (7.19). Thus q takes on the form

$$q = \frac{\exp(-\Delta G_{AU}/RT)}{\sum\limits_{N=A,U,G,C} \exp(-\Delta G_{NU}/RT)} \tag{7.22}$$

which becomes, after rearrangement,

$$q = \frac{1}{1 + \sum\limits_{N=G,C,U} \exp\left[(\Delta G_{AU} - \Delta G_{NU})/RT\right]} \tag{7.23}$$

It is at once apparent that q is unity only under limiting conditions:

$$\lim_{(\Delta G_{AU} - \Delta G_{NU}) \to \infty} q = 1 \tag{7.24}$$

or

$$\lim_{T \to 0} q = 1 \tag{7.25}$$

Thus as long as the ΔG values are finite and T is greater than zero, q and therefore Q are less than unity. It is obvious that this conclusion applies to all bases in the chain. This confirms our earlier, qualitative statement that the finite interaction energies and the thermal motion of molecules always prevent absolute accuracy in the recognition process.

7.4. The Physics of Complementary Base Recognition

The mechanism of the template-instructed synthesis of nucleic acids is based principally upon the formation of specific patterns of hydrogen bonding between complementary nucleic-acid bases. As Fig. 2.3 shows, the base pairs resulting from these interactions are geometrically almost identical. However, non-complementary bases can also interact with one another, and this can lead to the formation of base pairs in violation of the strict rules of complementarity.

What causes the difference between complementary and non-complementary base pairs and what makes the former so particularly suitable for the process of molecular recognition? One difference lies in the fact that the non-complementary base pairs are sterically quite different from each other and from the isomorphic, complementary pairs AU and GC (Fig. 7.3). But there are also considerable differences in stability between complementary and non-complementary base pairs (Table 7.1). For example, an AU pair is ten times stabler than an AA or a UU pair. The higher stability of the complementary base pairs in turn allows a higher fidelity of replication, as shown in Section 7.3.

The stability data summarized in Table 7.1 are the results of dielectric measurements in apolar solvents. In water, however, the individual base pairs are so unstable that it is not possible to detect their existence by normal methods. The reason for this is that the polar H_2O molecules form hydrogen bonds with the polar groups (N, NH, NH_2, OH and CO) of the organic bases in the nucleic acid molecule. These hydrogen bonds compete with those which hold the base pair together, so that pairing of single bases does not occur in aqueous solvents. The use of non-aqueous solvents for measurements of the constants of interaction shown in Table 7.1 means that only the *relative* stabilities of the various base pairs can be assessed. Experiments have shown that all base pairs are formed very rapidly; the magnitude of the rate constants implies that almost every collision between two nucleotides leads to base-pair formation.

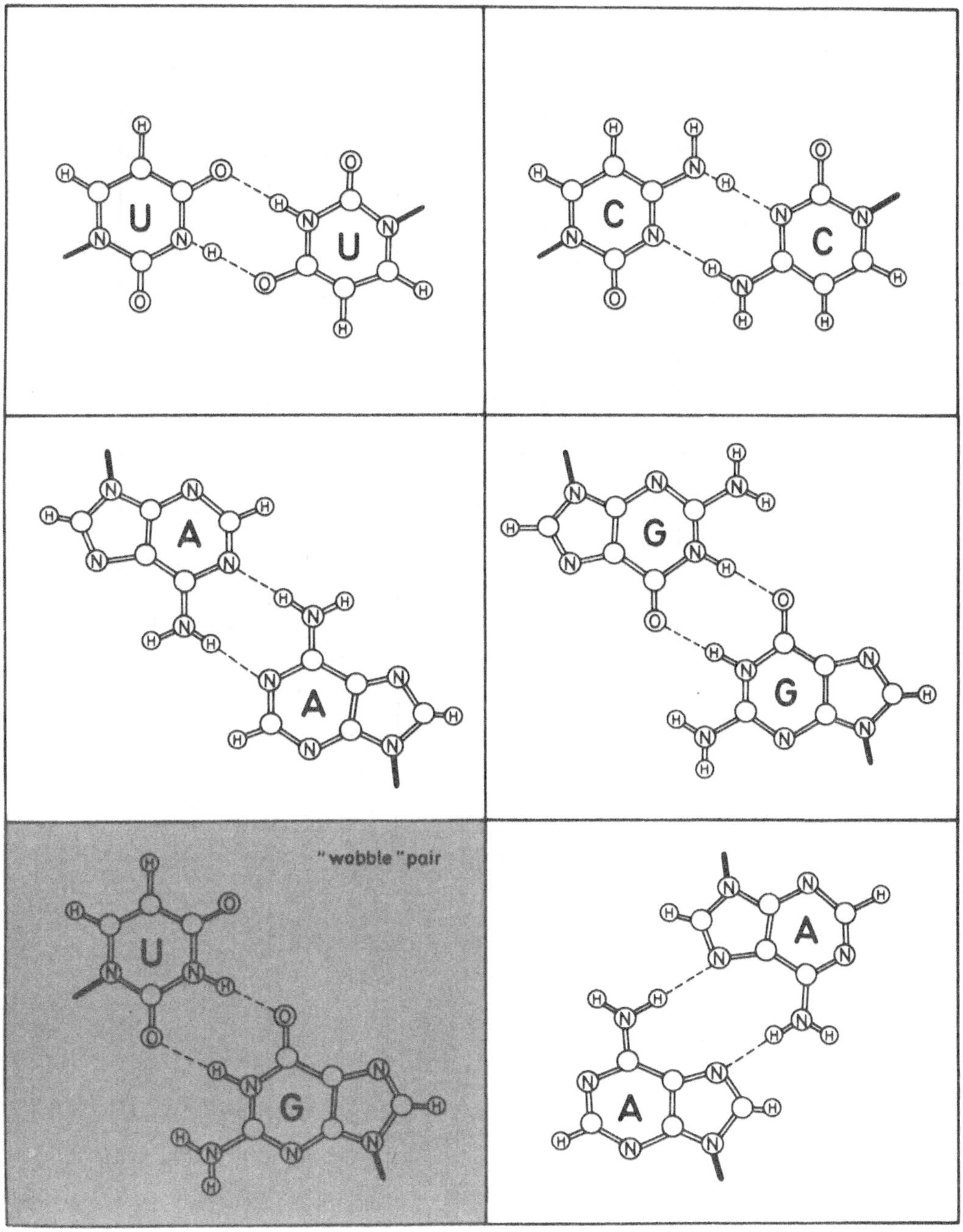

Fig. 7.3. Non-complementary base pairs.

CCl_4 / C_6H_6	U	A	C	G
U	45 / 15	550	<50	$<10^3$
A	150	22 / 8	<50	$<10^3$
C	<28	<28	50 / 28	$>10^4$
G	$<1.2 \cdot 10^3$	$<1.2 \cdot 10^3$	$3 \cdot 10^4$	$\approx 10^3$ / $1.2 \cdot 10^3$

Table 7.1. Stability constants of various base pairs in non-polar media. The table shows the association constants measured at 25°C, using dielectric methods, in non-polar solvents. Since some of the base combinations can pair in various ways the association constants represent in each case the sum of all such possibilities. (Measurements by T. Funck and co-workers in the Max Planck Institute for Biophysical Chemistry, summarized in Eigen, 1971; Kyogoku *et al.*, 1969.)

From the above discussion it is clear that the accurate transmission of genetic information requires special mechanisms to improve the process of single symbol recognition. Such mechanisms have been found and intensively investigated in recent years. They are based upon so-called *co-operative* phenomena, which are commonly observed among biological macromolecules. A relatively easily comprehensible exposition of co-operative effects has been presented by J. Engel and G. Schwarz (1970). The most important results relevant to the transmission of genetic information in nucleic acids will now be set out.

Let us consider a nucleic acid composed of a single nucleotide type such as the homopolymer poly(A):

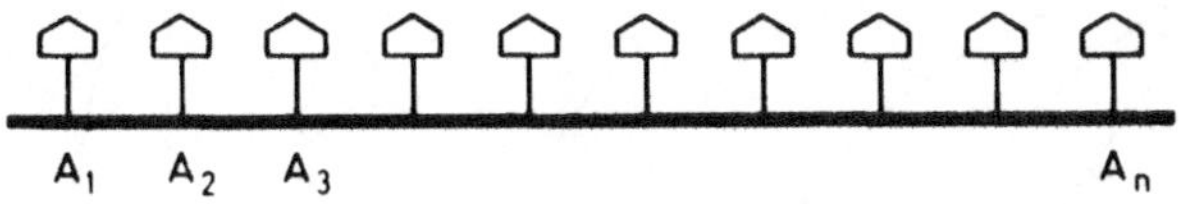

The synthesis of a daughter nucleic acid, instructed by this template, must obey the principle of complementary base recognition. During the replication process the intermediate

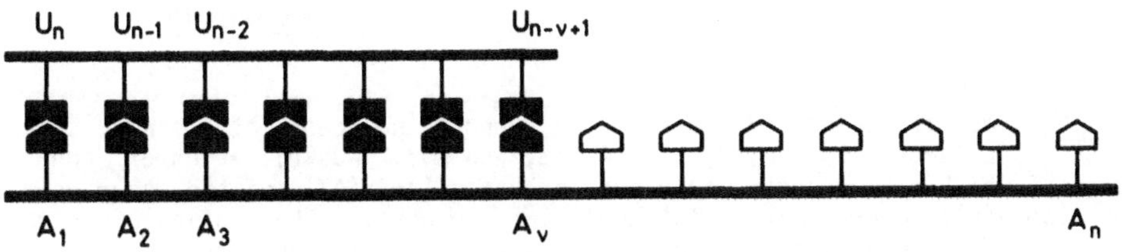

appears. We consider now the double-stranded segment and shall attempt to argue back from the stability of the whole segment to that of a single base pair. In other words, we wish to find out how much the co-operative effect of a chain of v base pairs affects the stability of a single member of the chain. The existence of such co-operative effects is demonstrated in Fig. 7.4. If temperature of a double helix, containing *e.g.* A and U bases, is slowly raised then the individual base pairs become unstable and the structure begins to "melt". In the high-temperature limit all the base pairs have dissociated and we are left with the corresponding single-stranded nucleic acid molecules in a randomly coiled state. But the transition does not occur gradually: it shows co-operative behaviour, that is, all the base pairs dissociate more or less at the same time. As the chain length increases, we note that (a) the region of temperature where the transition takes place rises, and (b) the curve shows an increasingly sharp transition.

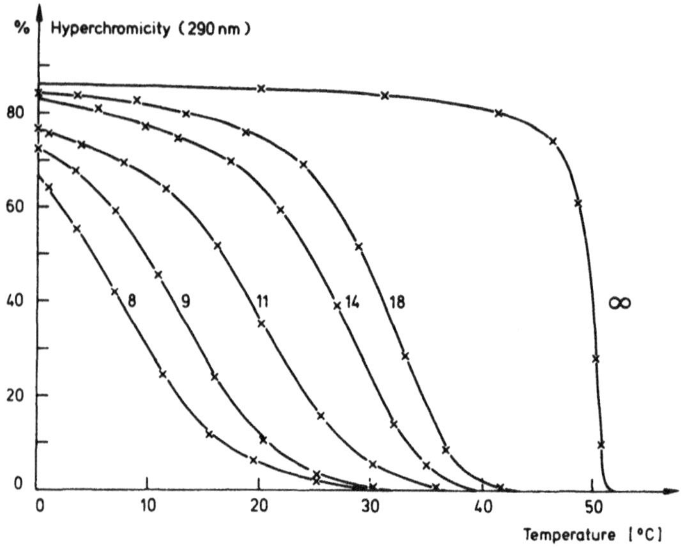

Fig. 7.4. Temperature-induced melting of double helices into single strands for various chain lengths (ν = 8, 9, 11, 14, 18, and ∞). Nucleic acids of equal chain length, of which one possessed only adenine bases (poly(A)) and the other only uracil bases (poly(U)), were mixed in equimolar amounts. The ratio of double- to single-stranded molecules is shown as a function of temperature: the observable quantity is here the increase in ultra-violet absorption (hyperchromicity) upon formation of helices. (From Pörschke, 1977.)

Let us attempt to interpret these results by formalizing the process of base-pairing. For this purpose we shall distinguish only two states of a base pair:

(1) nucleotides in an unpaired state (N), symbolized by 

(2) nucleotides in a paired state (N^*), symbolized by 

Among the 2^ν possible states of a nucleic acid of chain length ν there are accordingly two extreme forms, the idealized single strand

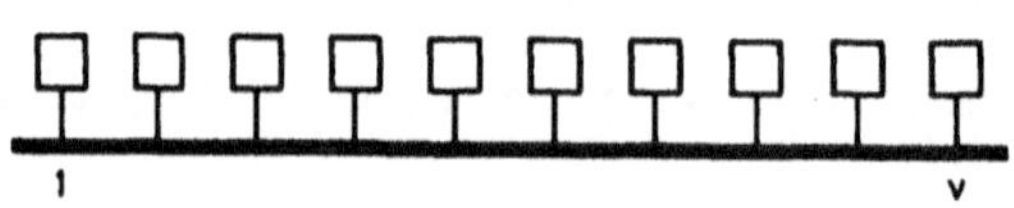

and the idealized double strand

Between these there are further $(2^{\nu-2}-2)$ intermediate arrangements of the form

A double helix for which $\nu = 100$ thus has $2^{100} \approx 10^{30}$ conceivable states of base-pairing. The base-pairing process, that is, the transition

will in general be influenced in a variety of ways by co-operative interaction between adjacent members of the chain. We shall make the simplifying assumption that the state of a single unit in the chain is influenced only by the state of its two immediate neighbours. This assumption is equivalent to the model of a linear lattice put forward by E. Ising (1925), in the context of a theory of ferromagnetism, and analysed quantitatively by H.A. Kramers and G.W. Wannier (1941).

In biophysics the Ising model was first applied by B.H. Zimm and J.K. Bragg (1958, 1959) to the problem of the helix-coil transition in polypeptides. Their results can be applied directly to double helix formation in nucleic acids. It allows the description of co-operative phenomena (in the linear model) by two basic thermodynamic parameters, an equilibrium constant s and a co-operativity parameter β :

(1) EQUILIBRIUM CONSTANT. The lengthening of sequences of N- or N^*-states

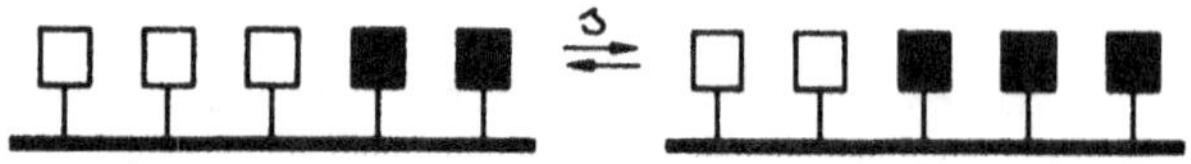

is described by a parameter s, which is the equilibrium constant of this transition, *i.e.*,

$$s = \frac{\bar{c}_{NN^*N^*}}{\bar{c}_{NNN^*}} \tag{7.26}$$

where $\bar{c}$ is the equilibrium concentration.

(2) CO-OPERATIVITY PARAMETER. In contrast to the transition described in (1), the process

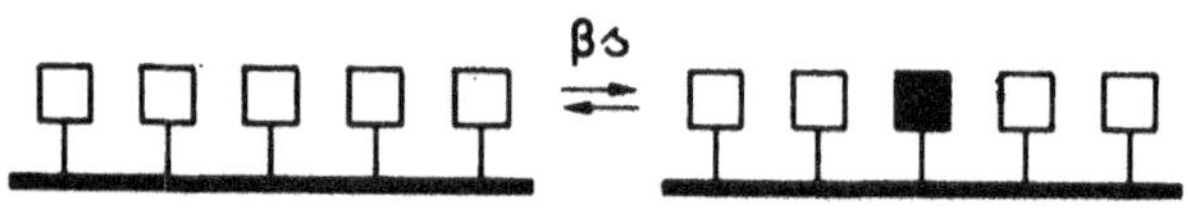

represents the nucleation or annihilation of an N^*-sequence. The equilibrium constant for this is

$$s = \frac{\bar{c}_{NN^*N}}{\bar{c}_{NNN}} \tag{7.27}$$

The parameter β clearly represents a measure of the co-operativity of the transition:

(a) For $\beta < 1$ the units in the chain tend to adopt the state of their neigbours (*positive co-operativity*).

(b) For $\beta = 1$ the equilibrium constant of a unit in the chain is no longer affected by the state of its neighbours (*no co-operativity*).

(c) For $\beta > 1$ neighbouring units tend to adopt different states (*negative co-operativity*).

The terminal units (N_1 and N_ν) are different from the others in that they each have only one direct neighbour. However, to simplify our model we shall assume that even at the ends of a polymer chain the nucleation process is described by the parameter β, and we note that, since our model does not consider the separation of strands after dissociation, all the equilibrium constants are dimensionless.

Let us illustrate the model with an example. The stages in the transition for an oligonucleotide with four chain units are (for low values of β):

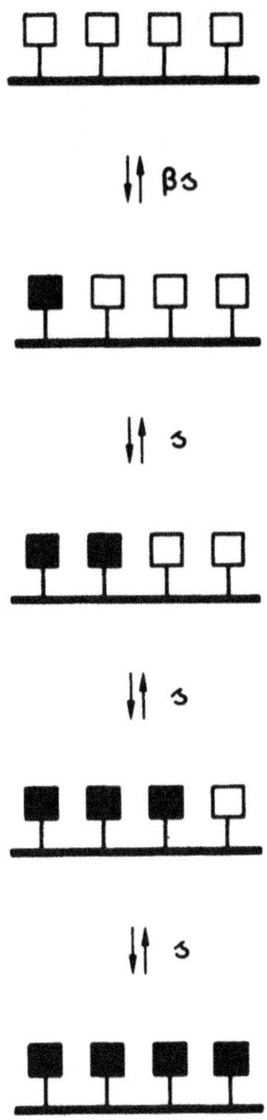

Fig. 7.5. Co-operative model for double helix formation. The stability of the first base pair (βs) is about 10^{-3} times that of subsequent, neighbouring base pairs. This increased stability can be explained by the existence of a "stacking" interaction (⇕) - each new base pair formed lies flat upon the last one in an energetically favourable way. (After Engel and Schwarz, 1970.) ▶

The equilibrium constant K_4 between the two extreme states is given by:

$$K_4 = \frac{\bar{c}_{N*N*N*N*}}{\bar{c}_{NNNN}} = \frac{\bar{c}_{N*N*N*N*}}{\bar{c}_{N*N*N*N}} \cdot \frac{\bar{c}_{N*N*N*N}}{\bar{c}_{N*N*NN}} \cdot \frac{\bar{c}_{N*N*NN}}{\bar{c}_{N*NNN}} \cdot \frac{\bar{c}_{N*NNN}}{\bar{c}_{NNNN}} = s^4 \tag{7.28}$$

If $K_4 = 1$ then the extreme forms NNNN and $N^*N^*N^*N^*$ are present in equal concentration. For a given K_4 value the concentrations of the intermediates depend solely upon the co-operativity parameter β. This can be clarified with a numerical example. For $\beta = 10^{-16}$ and $K_4 = 1$, Eq. (7.28) leads to the following values for the equilibrium constants of the intermediate forms:

$$K_1 = 10^4,\ k_2 = 10^8,\ K_3 = 10^{12} \tag{7.29}$$

and for their equilibrium concentrations:

$$\begin{aligned} \bar{c}_{N*N*N*N} &= 10^{-4}\ \bar{c}_{NNNN} \\ \bar{c}_{N*N*NN} &= 10^{-8}\ \bar{c}_{NNNN} \\ \bar{c}_{N*NNN} &= 10^{-12}\ \bar{c}_{NNNN} \end{aligned} \tag{7.30}$$

where

$$\bar{c}_{NNNN} = \bar{c}_{N*N*N*N*} \tag{7.31}$$

In the limiting case when $\beta \rightarrow 0$, the concentrations of the intermediate forms tend to zero and the equilibrium mixture consists solely of the pure N- and N^*-forms.

The example discussed here can easily be generalized to cover nucleic acids with ν chain members. The equilibrium constant is:

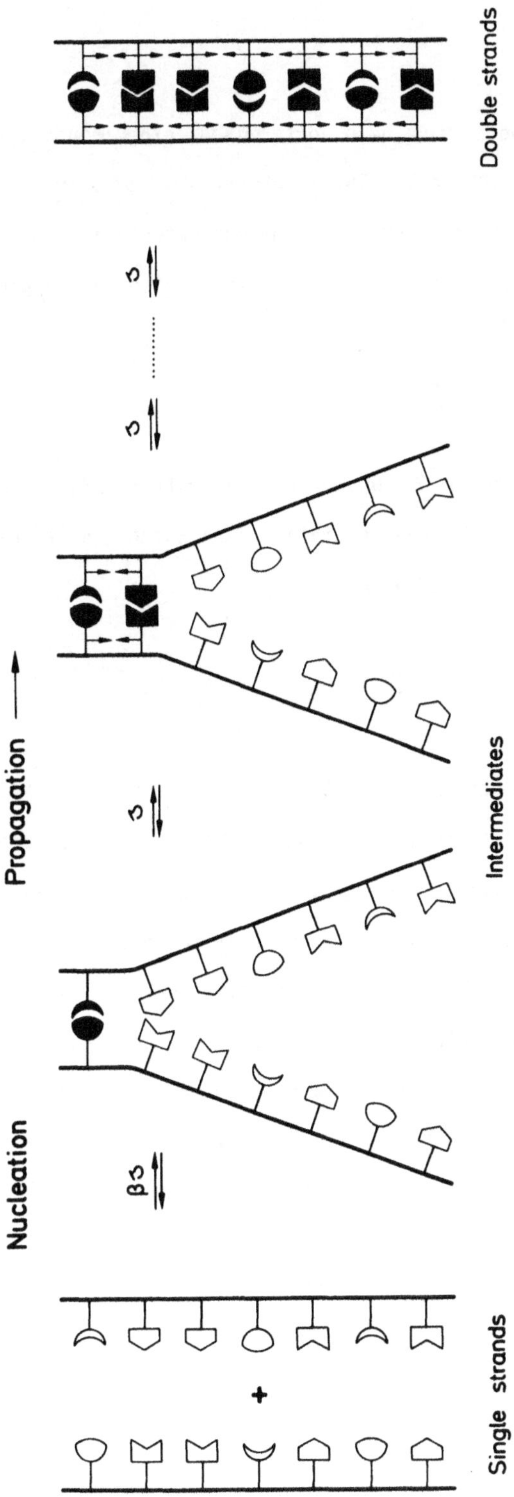
Double strands
σ
σ
Propagation
Intermediates
σ
Nucleation
βσ
Single strands
+

$$K_\nu = \beta s^\nu \qquad (7.32)$$

The parameters β and s have been determined experimentally for the complementary base pairs AU and GC (Pörschke, 1971; Pörschke *et al.*, 1973; Tinoco *et al.*, 1973). For homogeneous sequences of AU pairs the parameter s is about ten times smaller than for homogeneous GC pair sequences. The actual values are approximately:

$$s_{AU} \approx 10 \quad \text{and} \quad s_{GC} \approx 100 \qquad (7.33)$$

By contrast, the parameter β, which in real systems reflects not only co-operativity but also the entropy of strand separation, and which therefore has dimensions, is about the same for AU and GC pairs:

$$\beta \approx 10^{-3}\ M^{-1} \qquad (7.34)$$

While for AU pairs β has been directly measured, for GC pairs it had to be determined indirectly. It is clear from the value of β that we are dealing with a process showing strong, positive co-operativity. The first base pair has about one thousandth the strength of those subsequently formed.

Where is the cause of this enormous difference in stability to be sought? There is a stabilizing force between the bases stacked one upon another (see Fig. 7.5; *cf.* Section 2.1). The first, isolated base pair is handicapped by the fact that it can be held together only by hydrogen bonds, which, as we already know, hardly stabilize a base pair in aqueous solution. Only when another (or several other) base pairs have been added does the system manifest a co-operative stabilization. This stacking effect is particularly important in aqueous solution, and is presumed to be a result of hydrophobic and dipole-dipole interactions between the aromatic ring systems of the bases.

7.5. Information Storage in Darwinian Systems

The essential quantities determinating q (*cf.* Eq. 7.20) are the stability constants for base-pairing ($s_{N_+N_-}$) and the monomer concentrations (m_N). We shall now attempt to estimate the magnitude of the accuracy q of single symbol transmission, on the basis of the experimental data available. The co-operative model described in the foregoing section has provided us with experimental values for the stability constants of complementary base pairs in aqueous solution (Eq. 7.33). It should be noted that the values of these equilibrium constants depend upon such variables as temperature, salt concentration and so forth. (The s_{AU} and s_{GC} values given here were measured at a relatively low temperature, $T \approx 0^{\circ}C$).

The stability constants of GU pairs ("wobble" pairs) have not yet proved accessible to measurement. Experimental data however indicate indirectly that these constants are of similar magnitude to those of AU pairs, *i.e.*,

$$s_{GU} \approx s_{UG} \approx s_{AU} \approx s_{UA} \approx 10 \tag{7.35}$$

For the remaining stability constants there are at present no reliable experimental values at all, so that we must use current estimates of the experimentally undetermined quantities in Eq. (7.19). Thus the value of the constant s_{AC} is thought to lie between 1 and 2, while the values of s_{RR} and s_{YY} are probably below 1 (R = purine base, Y = pyrimidine base).

Another experimentally undetermined quantity, which renders the determination of q uncertain, is the monomer concentration m_N, even though Eq. (7.19) admittedly only calls for relative values of monomer concentrations. To see what influence m_N has upon q, Table 7.2 gives the q-parameters resulting from two different monomer concentrations. We should add that it is certainly unrealistic to imagine a uniform distribution of these in the primordial environment. The resulting q-values therefore represent an optimal, upper limit. Enzyme-free RNA replication cannot proceed with an accuracy of reproduction above 0.99.

Concentration of monomers	Stability constants of base pairs	Accuracy q	
		GC	AU
$m_A = m_U = m_G = m_C$	$s_{RR}=s_{YY}=1$ $s_{AC}=1$, $s_{GU}=10$ $s_{AU}=10$, $s_{GC}=100$	0.93	0.59
	$s_{RR}=s_{YY}<<1$ $s_{AC}=1$, $s_{GU}=5$ $s_{AU}=10$, $s_{GC}=100$	0.97	0.78
$m_U <[m_G, m_C]< m_A$	$s_{RR}=s_{YY}=1$ $s_{AC}=2$, $s_{GU}=10$ $s_{AU}=10$, $s_{GC}=100$	0.86	0.25
e.g. $m_U = \frac{1}{100} m_A$ $m_G=m_C = \frac{1}{10} m_A$	$s_{RR}=s_{YY}<<1$ $s_{AC}=1$, $s_{GU}=5$ $s_{AU}=10$, $s_{GC}=100$	0.95	0.69

Table 7.2. Accuracy of complementary base recognition in dependence on the monomer concentration and the stability constants of various base-pair combinations. (After Eigen and Schuster, 1979.)

What are the consequences of this result with regard to the quantity of information which can be reproducibly transmitted in a Darwinian system? This question is unambiguously answered by the threshold relationship (7.16). Let us first consider the superiority parameter Θ_m. We know little about this quantity, which depends on various selection-kinetic parameters (such as the growth and decay parameters A_k and D_k). However, it is the logarithm of this quantity which occurs in the threshold relation, and small changes in Θ_m therefore have little effect on ν_{max}. E. Domingo *et al.* (1978) have carried out a series of experiments with Q_β phage RNA which allow the conclusion (*cf.* Eigen and Schuster, 1979) that in this case the differential advantage lies in the region $\Theta_m = 4$ to $\Theta_m = 20$. Allowing for the uncertainty in Θ_m and in the error rate $(1-\bar{q})$, the maximum information capacity ν_{max} agrees closely with the actual information content.

It is apparent from Table 7.3 that enzyme-free RNA replication cannot reproduce more than 100 nucleotides under selection pressure without error. This chain length is approximately that of a tRNA molecule of today ($75 < \nu_{tRNA} < 90$). Higher accuracy of transmission is therefore only possible with the aid of optimized enzymes of replication. Studies *in vitro* of the RNA replication system of the virus Q_β (*cf.* Batschelet *et al.*, 1976) imply an accuracy of transmission of single symbols exceeding 0.9997. This permits the correct copying of about 3000 nucleotides, which corresponds roughly to the natural chain length of the Q_β phage RNA (≈ 4500 nucleotides). The increase in accuracy by a factor of over 100 in comparison with the enzyme-free process is probably to be attributed to the exact fitting of single nucleic-acid bases demanded by the enzyme's active site.

A further increase in the accuracy $\bar{q}$, and thus in the information capacity ν_{max}, is achieved in bacteria, by means of enzymes with a double function; in addition to each replication step they carry out "proof-reading" of the newly incorporated nucleotide. This allows the storage of about 10^6 molecular symbols in a bacterial genome.

To overcome the information threshold of 10^6 symbols, still more new mechanisms, such as genetic recombination, are necessary. Eucaryotic DNA reproduction raises the storage capacity to over 10^9 nucleotides, which is the amount of information encoded in the genome of a human being.

The threshold values just discussed were based upon more or less rough estimates of the reproductive accuracy $\bar{q}$ and the superiority parameter Θ_m to be expected in a natural population. Details of the reasoning behind these estimates are given by M. Eigen and P. Schuster (1979). In spite of the approximations, one aspect of the

Biological entity	Molecular mechanism	Error rate $(1-\bar{q})$	Superiority Θ_m	Greatest chain length ν_{max}	Examples in Nature
Prebiotic RNA molecules	Enzyme-free RNA replication *	$5\cdot10^{-2}$	2 20 $2\cdot10^{2}$	14 60 106	tRNA precursors ($\nu \approx 80$)
Viruses	RNA-dependent synthesis by specific replicases	$5\cdot10^{-4}$	2 20 $2\cdot10^{2}$	1386 5991 10597	Q_β phage RNA ($\nu \approx 4500$)
Bacteria	DNA replication by polymerases and correction enzymes	$1\cdot10^{-6}$	2 20 $2\cdot10^{2}$	$0.7\cdot10^{6}$ $3.0\cdot10^{6}$ $5.3\cdot10^{6}$	*E.coli* bacterium ($\nu \approx 4\cdot10^{6}$)
Vertebrates	DNA replication and recombination in eucaryotic cells	$1\cdot10^{-9}$	2 20 $2\cdot10^{2}$	$0.7\cdot10^{9}$ $3.0\cdot10^{9}$ $5.3\cdot10^{9}$	*Homo sapiens* ($\nu \approx 3\cdot10^{9}$)

◀ Table 7.3. Error rates of single symbol copying and greatest chain lengths predicted by the threshold relation (7.16).

*Enzyme-free RNA replication has not yet been observed to a significant extent under normal laboratory conditions. It is however known that such processes can occur on catalytically active surfaces (minerals: silicates, phosphates *etc.*) and in the presence of certain metal ions (Bridson and Orgel, 1980; Lohrmann *et al.*, 1980). It is reasonable to assume, whatever the mechanism, that the accuracy of the enzyme-free RNA replication is determined by the ratios between the binding constants of correct and incorrect nucleotides. The error rate calculated from the experimental data at present available (Table 7.1) is approximately $5 \cdot 10^{-2}$. (After Eigen and Schuster, 1979.)

origin of information becomes particularly clear: *a significant expansion of information capacity demands a significant improvement in the accuracy with which single symbols are copied.* The decisive step in the origin of life was therefore without doubt the transition from enzyme-free to enzyme-dependent replication of RNA, since, as the above analysis shows, without the help of complex molecular copying-machines (polymerases) an average of not more than 100 nucleotides in a nucleic acid can be correctly copied under selection pressure. On the other hand, we know that the blueprint for such polymerases certainly requires more than 100 symbols and could therefore only have arisen by a process of selective optimization of the nucleic acids. This quandary is obviously the molecular-genetic version of the well-known problem of the chicken and the egg: which came first? The nucleic acids or the proteins? This important question will occupy the centre of our discussion to come.

8. Self-Organization in Macromolecular Networks

As shown above, prebiotic conditions of nucleic acid replication set certain limits to the storage capacity of single information carriers. In the absence of enzymes the greatest quantity of information which can be transmitted reproducibly is far too small for the construction of an integrated genetic system such as the simplest organisms known today. We must therefore ask whether there are conceivable information-processing systems which, working under prebiotic conditions, could in spite of their high error rate lead to a progressive evolution. Up to now we have considered only self-organization processes in which molecular species reproduce independently and therefore compete with each other. We shall now extend our theory so as to include interactions between the molecular species which stabilize their genetic information by co-operative reproduction and co-evolution.

Needless to say, it is an impossible task to analyse every conceivable network of interactions. Our goal will be to find the logically simplest form of organization which has the property of stabilizing information and which leads to the emergence of an autonomously reproducing genetic system comparable to a living cell.

8.1. General Selection Equations

We shall ultimately attempt to analyse the dynamics of self-organization processes in cyclic and non-cyclic reaction networks. For this purpose we first formulate equations which describe in a general way the growth of self-organizing systems, and shall then classify them according to reaction mechanism.

Consider again the model system for molecular self-organization depicted in Fig. 4.1. Without constraints, the most general form of the differential equations for the concentration variables x_i is

$$\dot{x}_i = \Gamma_i(\mathbf{x}, \mathbf{k}, \mathbf{x}(0)) - \varphi x_i \qquad (i = 1, \ldots\ldots, N) \tag{8.1}$$

where Γ_i is a growth function and φ a dilution parameter, which is in turn related to the global dilution flux Φ_0 by (*cf.* Eq. 5.8):

$$\varphi = \frac{\Phi_0}{\sum_j x_j} \tag{8.2}$$

The general growth function Γ_i depends upon $\mathbf{x}$, $\mathbf{k}$ and $\mathbf{x}(0)$. Thus the vector

$$\mathbf{x} = \begin{pmatrix} x_1 \\ \vdots \\ x_N \end{pmatrix} \tag{8.3}$$

expresses all the concentration variables x_j, while the vector

$$\mathbf{k} = \begin{pmatrix} k_1 \\ \vdots \\ k_j \end{pmatrix} \tag{8.4}$$

summarizes all the constant kinetic parameters k_j (rate constants, equilibrium constants, concentration of energy-rich monomers *etc.*). Finally the components of the vector

$$\mathbf{x}(0) = \begin{pmatrix} x_1(0) \\ \vdots \\ x_N(0) \end{pmatrix} \tag{8.5}$$

are the values of the variables x_i at time t=0, *i.e.*, are the starting conditions of Eq. (8.1).

The change of the total concentration c with time is given by the following differential equation:

$$\dot{c} = \sum_i \dot{x}_i = \sum_i \Gamma_i(\mathbf{x}, \mathbf{k}, \mathbf{x}(0)) - \Phi_0 \tag{8.6}$$

The dilution flux Φ_0 again allows us to regulate the total concentration in the system and to fix it at any desired value. Thus the total concentration in the system is constant ($\dot{c}=0$) if at every moment the dilution flux Φ_0 exactly cancels out the population growth, *i.e.*, when

$$\Phi_0 \equiv \sum_i \Gamma_i(\mathbf{x}, \mathbf{k}, \mathbf{x}(0)) \tag{8.7}$$

Condition (8.7) is clearly sufficient to establish the familiar CP constraint. In this case the x_i variables are given by the differential equations

$$\dot{x}_i = \Gamma_i(\mathbf{x}, \mathbf{k}, \mathbf{x}(0)) - \frac{x_i}{c_0} \sum_i \Gamma_j(\mathbf{x}, \mathbf{k}, \mathbf{x}(0)) \qquad (i = 1, \ldots\ldots, N) \tag{8.8}$$

where c_0 is the stationary concentration of all information carriers in the CP system. Equations (8.8) are the generalized equations for the dynamics of self-organizing molecular systems under CP conditions.

We can now evaluate the selection equations (8.8) for any given reaction mechanism by discussing various types of *Ansatz* for Γ_i. In the special case of non-coupled information carriers, for example, with

$$\Gamma_i = W_i^{CP} x_i + \sum_{j \neq i} \psi_{ij} x_j \tag{8.9}$$

and

$$\frac{1}{c} \sum_i \Gamma_i = \bar{E} \tag{8.10}$$

we obtain the familiar CP selection equations (5.20) (see Section 5.2). Here the relation (8.10) is equivalent to the mass-balance condition (5.19).

Without yet going into details of reaction mechanism, we can already specify the function Γ_i using general criteria of reaction kinetics. If we write out the growth function as a sum of contributions of zeroth, first, second and higher order,

$$\Gamma_i = k_i + \sum_j k_{ij}x_j + \sum_j \sum_k k_{ijk}x_j x_k + \ldots\ldots\ldots \tag{8.11}$$

then we can define the *degree* p_i of the growth function Γ_i as the highest power appearing in Eq. (8.11). For example, the growth function given in Eq. (8.9) has the degree p=1. A system of N species can thus have a spectrum of p-values. If these are all identical,

$$p_1 = \ldots\ldots = p_N \equiv p \tag{8.12}$$

we call the system *homogeneous*. If condition (8.12) is not met, the system is *inhomogeneous* with regard to its growth functions.

In the treatment which follows we shall restrict consideration of self-organization processes in macromolecular networks to those under CP constraint. This includes *all* processes whose dynamics are described by Eqs. (8.8). We shall however see that our conclusions are basically independent of the form of the selection pressure, and that they can be extended directly to systems which are not under CP constraint. For this reason it will prove helpful to describe the self-organization process using normalized population variables.

We first introduce some notation. The composition of the system at any time t is given by an array of N numbers

$$\mathbf{x}(t) = \begin{pmatrix} x_1(t) \\ \vdots \\ x_N(t) \end{pmatrix} \tag{8.13}$$

and can therefore be represented by a point in N-dimensional concentration space $\mathbb{X}^N$, defined by

$$\mathbb{X}^N\colon \{x_1,\ldots\ldots,x_N \mid x_i \geq 0 \qquad (i = 1,\ldots\ldots,N)\} \tag{8.14}$$

If the molecular composition of the system changes, this point describes a trajectory

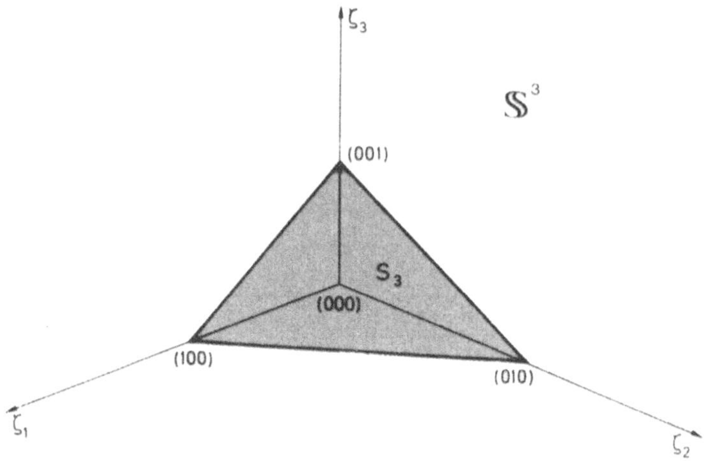

Fig. 8.1. Fitting S_3 into $\mathbb{S}^3$. The unit simplex S_N represents an (N-1)-dimensional hypersurface in N-dimensional space. The simplex representation is particularly suitable for the description of the long-term behaviour of self-organizing systems. Each point on the simplex represents a particular macromolecular composition of the system. For example a point at an apex of the simplex means that in the system only one species is present. The region outside the simplex is physically meaningless.

in $\mathbb{X}^N$ We now normalize the concentration variables x_i by the condition

$$\zeta_i = \frac{1}{c(t)} x_i \tag{8.15}$$

The new variables $\zeta_i = \zeta_i(t)$ then occupy a space $\mathbb{S}^N$, defined by

$$\mathbb{S}^N : \{\zeta_1, \ldots\ldots, \zeta_N \mid 0 \leq \zeta_i \leq 1\} \tag{8.16}$$

As a consequence of the normalization procedure the N degrees of freedom of the x_i-variables are reduced to N-1 degrees of freedom of the ζ_i-variables. Furthermore it is clear that the point in $\mathbb{S}^N$ describing the composition of the system has a locus defined by an (N-1)-dimensional surface S_N

$$S_N : \{\zeta_1, \ldots\ldots, \zeta_N \mid 0 \leq \zeta_i \leq 1 \quad \text{and} \quad \sum_i \zeta_i = 1\} \tag{8.17}$$

which is called a *simplex* or a *unit simplex*. The simplex S_N has the form of a regular polyhedron with N corners, of whose co-ordinates one is unity and the rest are zero. It is shown for N=3 in Fig. 8.1. The simplex representation can now be used to describe in a diagrammatic way changes in the macromolecular composition of a self-organizing model system.

8.2. Methods of Stability Analysis: Fixed Point Analysis

The differential equations which describe the network dynamics of self-reproducing information units are usually very complicated and soluble only by numerical methods. However, in many cases the point of interest is not the detailed time-course of the solutions but rather the long-term behaviour of the system. In our case, the final state is at issue: the nature and the dynamic stability of this final state determine whether or not different information carriers can co-exist and stabilize each other.

A suitable procedure for extracting information about the long-term behaviour of a dynamic system is the method of fixed point analysis (see Hirsch and Smale, 1974).

"Fixed points" are the states of a dynamic system that are invariant in time. In our model system, the fixed points are those points in N-dimensional concentration space $\mathbb{X}^N$ for which the macromolecular composition of the model system no longer changes. The positions of all the fixed points are accordingly determined by the stationary state condition

$$\dot{\mathbf{x}} = 0 \tag{8.18}$$

However, nothing in the real world is free of perturbations. We must therefore ask what macroscopic changes in the dynamic behaviour of our system appear when we introduce small perturbations in the concentration variables at given fixed points.

For small perturbations we can carry out a linear stability analysis, the principles of which are described in detail in Appendix A.2. This analysis will lead to a description of the dynamics of the perturbed system in terms of a spectrum of normal modes with reciprocal time constants ω_k; the latter appear as eigenvalues of a set of differential equations linearized around the fixed point. We have already made use of this procedure in Section 5.2. As already stated, the fixed points of a self-organizing system are its stationary points in concentration space. We shall now look at the topological aspects of this in more detail. There are four classes of fixed points, summarised in Table 8.1:

(1) STABLE FIXED POINTS or SINKS. All eigenvalues ω_k have negative real parts, so that all perturbations decay exponentially. The system reacts to fluctuations with an automatic compensation of the fluctuation. The Ehrenfest game discussed in Section 6.1 is a typical simulation of self-regulating fluctuations.

(2) SOURCES. All eigenvalues ω_k are real and positive, that is, all directions of fluctuation are unstable. Whenever a fluctuation appears, the system tends to reinforce it. The corresponding fixed point has the topological properties of a source, in that the system tends to move away from it.

(3) SADDLE-POINTS. One eigenvalue ω_i is real and positive and the other real and negative, so that one direction of fluctuation is unstable. Saddle-points thus behave as sources but only in particular directions.

(4) GENERAL UNSTABLE FIXED POINTS. Some eigenvalues ω_i are zero or pure imaginary. Systems with pure imaginary eigenvalues make up so-called centres, whose trajectories consist of a family of concentric circles.

The basic principle of fixed point analysis now becomes clear: the dynamics of the system are determined by the form of the potential surface in concentration space. This potential surface possesses in general a set of local maxima and local minima which correspond to the fixed points of the system. The set of fixed points thus represents a topographical abstraction of the most important structural features of the complete potential surface. If we now find the lines ("valleys") joining the individual points, then we can, without needing to describe the entire potential surface, make general statements about the dynamics of the system: its time-dependent development will as a rule follow the downhill course of water in a mountain range.

Fixed-point class	Symbol	Eigenvalues	Designation	
1	●	$\omega_1, \omega_2 < 0$	Node	Sinks
		$\omega_1 = \omega_2 < 0$	Focus	
		$\omega_1 = -a + ib$ $\omega_2 = -a - ib$ $a, b > 0$	Spiral	
2	○	$\omega_1, \omega_2 > 0$	Source	
3	⊕	$\omega_1 > 0, \omega_2 < 0$	Saddle	
4	⦶ ⊖ 	$\omega_1 > 0, \omega_2 = 0$ $\omega_1 = 0, \omega_2 < 0$ $\omega_1 = +ib, \omega_2 = -ib$	Centre	

Table 8.1. Fixed-point classes for a two-dimensional dynamic system

However, self-organizing systems can also follow more complicated trajectories. Thus a "stable" long-term behaviour pattern can show periodic changes ("limit cycles") or even aperiodic changes (*cf.* Figs. A.2.1 and A.2.2). Pictorially expressed, a "limit cycle" corresponds to a never-ending, circularly-flowing river. Trajectories of this kind are never seen in the comparatively simple-structured gravitational field of the earth, except in the imagination of an artist such as M.C. Escher (Fig. 8.2).

Fig. 8.2. M.C. Escher *Waterfall* (1961). Reproduced by kind permission of the Escher Foundation (Collection Haags Gemeentemuseum, The Hague).

Sinks, closed orbits and aperiodic lines are subsumed under the concept "attractor". The direct neighbourhood of an attractor is called a "basin". We can thus imagine any potential surface as composed of a multitude of such basins, each one centred on an attractor. The basins are separated from each other by potential walls. All points on the potential walls are unstable. Any small perturbation brings the system away from an unstable fixed point and sets it moving, along a direct or an indirect path, by way of further unstable fixed points, towards the attractor. Such a description applies to all of the selection processes discussed up to now (*e.g.* in Section 5.10).

8.3. Long-Term Behaviour of Self-Organizing Systems

As we have seen, the nucleic acids reproduce according to the principle of complementary base recognition, so that the overall process always involves the syntheses of (+) and (-) strands (Fig. 2.3). From now on we shall frequently represent a self-reproductive nucleic acid unit I by the symbol:

In a non-coupled system (Fig. 8.3a) the result of selection competition is always the survival of one class of information carrier, represented by a quasi-species. Since each mutant within a quasi-species competes with its master sequence, there are limits to the accumulation of biological information which are set by the length of the master sequence and its accuracy in reproduction (Section 7.2).

We shall now show that expansion of the information content is attainable if two or more quasi-species are in some way coupled to each other so that they can benefit from co-existence. In order to solve this problem we must first find out what type of coupling is needed.

For this purpose we investigate two kinds of coupling, as shown schematically in Figs. 8.3b and 8.3c, and ask which of the reaction networks shown might lead to

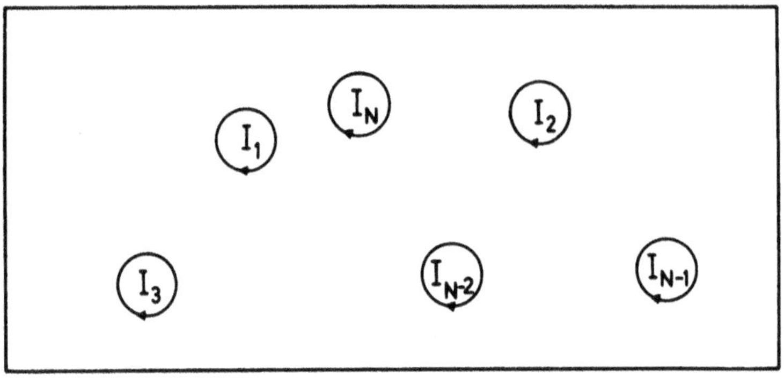

(a) Independently - reproducing information carriers

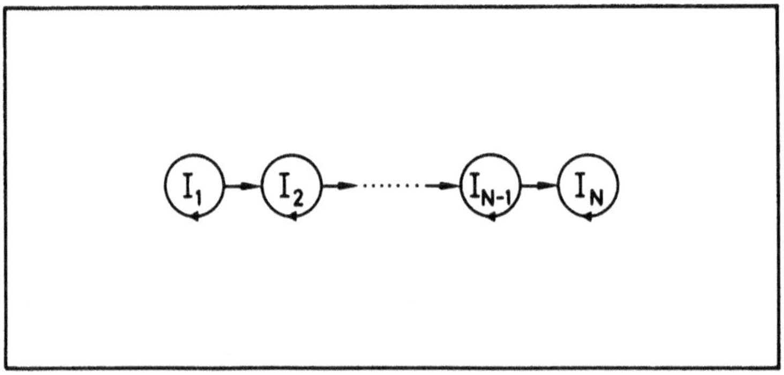

(b) Catalytic chain of information carriers

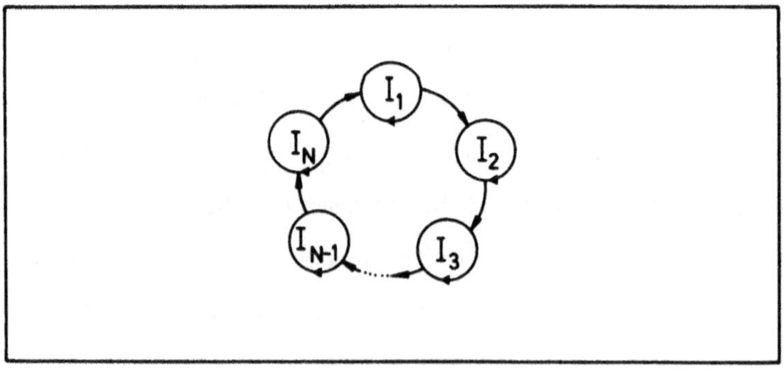

(c) Hypercycle

Fig. 8.3. Basic organization patterns of self-reproducing information carriers.

co-existence of the molecular species and thus to the stabilization of a larger amount of information than the error rate would allow for independently-growing quasi-species.

(1) INDEPENDENTLY-GROWING INFORMATION CARRIERS. We already know that the uncoupled system does not permit the co-existence of different information carriers. None the less, we shall consider this system in detail in order to illustrate the method of fixed-point analysis. To clarify the results (using the simplex representation), our analysis will be restricted to three-dimensional systems.

The reaction system specified in Fig. 8.3a is described by a linear growth function of the form

$$\Gamma_i = k_i x_i \tag{8.19}$$

This leads, applying Eq. (8.8), to the equation

$$\dot{x}_i = k_i x_i - \frac{x_i}{c_0} \sum_{j=1}^{3} k_j x_j \qquad (i = 1,2,3) \tag{8.20}$$

With the definition

$$\bar{k} = \frac{1}{c_0} \sum_j k_j x_j \tag{8.21}$$

we finally obtain

$$\dot{x}_i = (k_i - \bar{k})\, x_i \qquad (i = 1,2,3) \tag{8.22}$$

These differential equations have, as expected, the same structure as Eqs. (5.21) for the non-coupled reproduction of molecular species.

The long-term behaviour of the system is determined by the three fixed points

$$\mathbf{x}^{s_1} = \begin{pmatrix} c_0 \\ 0 \\ 0 \end{pmatrix}, \quad \mathbf{x}^{s_2} = \begin{pmatrix} 0 \\ c_0 \\ 0 \end{pmatrix}, \quad \mathbf{x}^{s_3} = \begin{pmatrix} 0 \\ 0 \\ c_0 \end{pmatrix} \tag{8.23}$$

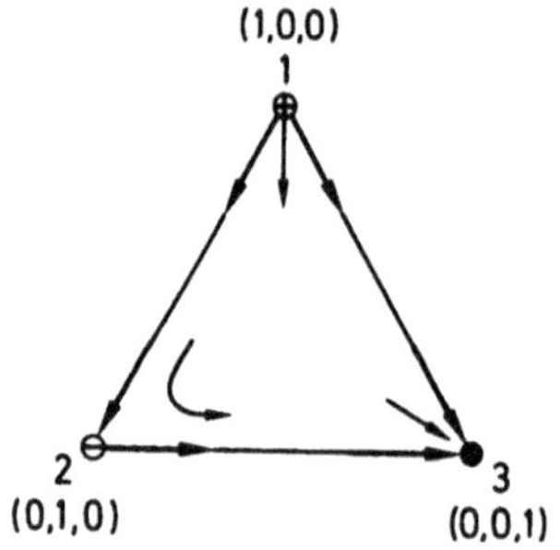

Fig. 8.4. Three-dimensional fixed-point map for independently replicating information carriers with parameter values $k_1=1$, $k_2=2$ and $k_3=3$. The system has only one stable fixed point (0,0,1). The fact that this fixed point lies at one corner of the concentration simplex is a criterion for the unambiguous selection of one of the three molecular species. The result is independent both of initial conditions and of the total concentration. (After Eigen and Schuster, 1979.)

which lie in the corners of the simplex S_3 (Fig. 8.4) and whose values have been calculated in Appendix A.2 for the corresponding n-dimensional case.

The result (see also Eq. 5.26) can be applied directly to the present case. Thus we obtain for each fixed point $\mathbf{x}^{S_i}$ a spectrum of three eigenvalues:

$$\begin{aligned} \omega_1^{(i)} &= k_1 - k_i \\ \omega_2^{(i)} &= k_2 - k_i \\ \omega_3^{(i)} &= - k_i \end{aligned} \tag{8.24}$$

If the parameters k_j are known then the topological properties of the three fixed points can be determined.

Figure 8.4 is based upon the values $k_1=1$, $k_2=2$ and $k_3=3$. The eigenvalue spectrum for the fixed point $\mathbf{x}^{S_1}$ is therefore:

$$\begin{aligned} \omega_1^{(1)} &= k_1 - k_1 = 0 \\ \omega_2^{(1)} &= k_2 - k_1 > 0 \\ \omega_3^{(1)} &= - k_1 < 0 \end{aligned} \tag{8.25}$$

That is, the fixed point $\mathbf{x}^{S_1}$ is topologically a saddle. For the fixed point $\mathbf{x}^{S_2}$ we have

$$\omega_1^{(2)} = k_1 - k_2 < 0$$

$$\omega_2^{(2)} = k_2 - k_2 = 0 \qquad (8.26)$$

$$\omega_3^{(2)} = - k_2 < 0$$

so that $\mathbf{x}^{S_2}$ corresponds to a centre. For the fixed point $\mathbf{x}^{S_3}$ we obtain:

$$\omega_1^{(3)} = k_1 - k_3 < 0$$

$$\omega_2^{(3)} = k_2 - k_3 < 0 \qquad (8.27)$$

$$\omega_3^{(3)} = -k_3 < 0$$

i.e., all the eigenvalues are negative, so that $\mathbf{x}^{S_3}$ is a sink (more precisely, a node, since the eigenvalues are different).

Thus of the three fixed points of the non-coupled system (Eq. 8.23), only one is stable, *viz.*, the one with the co-ordinates $x_1^{S_3}=0$, $x_2^{S_3}=0$ and $x_3^{S_3}=c_0$. This means that the co-existence of all three information units is impossible. Whatever the starting conditions, after a long enough time only species I_3 will survive, while I_1 and I_2 will have died out successively (see also Fig. 8.5).

In general, the co-existence of N information units is only given when at least one stable fixed point lies in the interior of the simplex covering the ζ_i variables. Non-rigorous selection, and thus partial co-existence, is the result when the fixed points lie on the edges, faces or hyperfaces of the concentration simplex.

(2) ACYCLIC CATALYTIC CHAINS OF INFORMATION CARRIERS. The simplest coupling between molecular species consists of a linear connection between successive units in a catalytic chain, as shown in Fig. 8.3b. In this type of coupling the growth of each information carrier is promoted by the previous one in the chain. Since the first unit in the chain has no predecessor, we must distinguish two classes of growth functions: for I_1 we have the simple, linear growth function

$$\Gamma_1 = k_1 x_1 \tag{8.28}$$

whereas the other information carriers grow according to a rate law containing a catalytic term expressing the promotion of growth of I_i by the species I_{i-1}:

$$\Gamma_i = k_i x_i + k_i' x_i x_{i-1} \qquad (i = 2,\ldots\ldots,N) \tag{8.29}$$

Thus in contrast to Eq. (8.28), the growth functions (8.29) are inhomogeneous.

The selection equations (8.8), with the above *Ansatz* for Γ, lead to the set of differential equations

$$\dot{x}_1 = k_1 x_1 - \frac{x_1}{c_0}\left[k_1 x_1 + \sum_{j=2}^{N} (k_j x_j + k_j' x_j x_{j-1})\right] \tag{8.30}$$

and

$$\dot{x}_i = k_i x_i + k_i' x_i x_{i-1} - \frac{x_i}{c_0}\left[k_1 x_1 + \sum_{j=2}^{N} (k_j x_j + k_j' x_j x_{j-1})\right] \qquad (i=2,\ldots\ldots,N) \tag{8.31}$$

For simplicity we shall again restrict our fixed-point analysis to the case of three dimensions. An elementary calculation gives six fixed points for Eqs. (8.30) and (8.31). The three fixed points

$$\boldsymbol{x}^{s_1} = \begin{pmatrix} c_0 \\ 0 \\ 0 \end{pmatrix}, \quad \boldsymbol{x}^{s_2} = \begin{pmatrix} 0 \\ c_0 \\ 0 \end{pmatrix}, \quad \boldsymbol{x}^{s_3} = \begin{pmatrix} 0 \\ 0 \\ c_0 \end{pmatrix} \tag{8.32}$$

lie in the corners of the simplex S_3 while the positions of the remaining three fixed points

$$\boldsymbol{x}^{s_4} = \begin{pmatrix} \dfrac{k_1-k_2}{k_2'} \\ \dfrac{k_2'c_0-k_1+k_2}{k_2'} \\ 0 \end{pmatrix}, \quad \boldsymbol{x}^{s_5} = \begin{pmatrix} 0 \\ \dfrac{k_2-k_3}{k_3'} \\ \dfrac{k_3'c_0-k_2+k_3}{k_3'} \end{pmatrix}, \quad \boldsymbol{x}^{s_6} = \begin{pmatrix} \dfrac{k_1-k_2}{k_2'} \\ \dfrac{k_1-k_3}{k_3'} \\ \kappa \end{pmatrix} \tag{8.33}$$

with

$$\kappa = \frac{k_2'k_3'c_0 - k_3'(k_1 - k_2) - k_2'(k_1 - k_3)}{k_2'k_3'} \tag{8.34}$$

depend linearly upon the parameter c_0.

As we have seen, the positions of the fixed points on the concentration simplex suffice to tell us the ultimate result of selection competition (*cf.* Fig. 8.5). The three fixed points $\mathbf{x}^{S_1}$, $\mathbf{x}^{S_2}$ and $\mathbf{x}^{S_3}$ are in the corners of the simplex S_3, so that in each case two of the three co-ordinates are zero-valued, and this in turn means - independently of the topological nature of the fixed points - that there is no co-existence of information units in these stationary states. The two fixed points $\mathbf{x}^{S_4}$ and $\mathbf{x}^{S_5}$, in dependence upon c_0, move along a trajectory defined by the simplex edges $\overline{12}$ and $\overline{23}$ respectively, so that in each case two out of the three information units can co-exist. Only fixed point $\mathbf{x}^{S_6}$ can, for particular c_0 values, move into the interior of the simplex, indicating co-existence of all three components. However, for this to occur the inequalities

$$k_1 > k_2, k_3 \tag{8.35}$$

and

$$c_0 > \frac{k_1 - k_2}{k_2'} + \frac{k_1 - k_3}{k_3'} \tag{8.36}$$

must hold.

These conditions follow from the requirement for values greater than zero for all co-ordinates of $\mathbf{x}^{S_6}$. The result can easily be extended to the N-dimensional case. The appropriate relations are

$$k_1 > k_j \qquad (\text{for all } j \geq 2) \tag{8.37}$$

Fig. 8.5. Three-dimensional fixed-point map for acyclic catalytic chains of information carriers. The dynamics of the system are described by the differential equations (8.30) and (8.31). In the example shown here the rate constants have the following values:

$$k_1=3,\ k_2=2,\ k_3=1 \text{ and } k_2'=2,\ k_3'=1$$

The positions of the individual fixed points then depend linearly upon the parameter c_0 (see Eqs. 8.32 and 8.33).

(a) At low concentration the fixed points $\mathbf{x}^{S_4}, \mathbf{x}^{S_5}$ and $\mathbf{x}^{S_6}$ lie outside the physically meaningful region in concentration space, *i.e.*, at least one co-ordinate of a fixed point is negative.

(b) If the total concentration attains the critical value

$$c_0 = c_{cr} = (k_1-k_2)k_2' + (k_2-k_3)k_3'$$

then the fixed point $\mathbf{x}^{S_6}$ reaches the $\overline{12}$ edge of the simplex S_3.

(c) If the concentration exceeds the critical value, c_{cr}, the fixed point $\mathbf{x}^{S_6}$ moves into the interior of the simplex. At the same time it changes its topological nature and becomes a spiral. The co-ordinates of $x_1^{S_6}$ and $x_2^{S_6}$ are constant, so that for high values of c_0 the system approaches the state $x_3^{S_6} = c_0$.

(d) For $c_0 \to \infty$, the fixed point $\mathbf{x}^{S_6}$ moves into corner 3 of the concentration simplex, which leads to the exclusive selection of co-ordinate $x_3^{S_6}$. In the corner of the simplex there is by definition no co-existence.

Topological details of the fixed points may be obtained from the eigenvalue spectrum. This can be obtained by the method described in Appendix A.2. (From Eigen and Schuster, 1979.)

and

$$c_0 > \sum_{j\geq 2} \frac{k_1 - k_j}{k_j'} \tag{8.38}$$

It is clear that the restrictions (8.37) and (8.38) impose very special requirements upon the system. For example, no mutant may appear in the three-component system which violates inequality (8.36). Even if a system could evolve and adhere to this restriction, the consequence would be a relative shift of several orders of magnitude in the population numbers of the co-operating information carriers, since only the last member of the reaction chain multiplies at rate proportional to c_0. This tendency would be highly disadvantageous for the development of a common function over any reasonable concentration range. We are thus forced to conclude that the acyclic coupling of information carriers will not usually result in the stabilization of their information. This applies not only to linear chains, as shown in Fig. 8.3b, but also to branched chains (see Eigen and Schuster, 1979.)

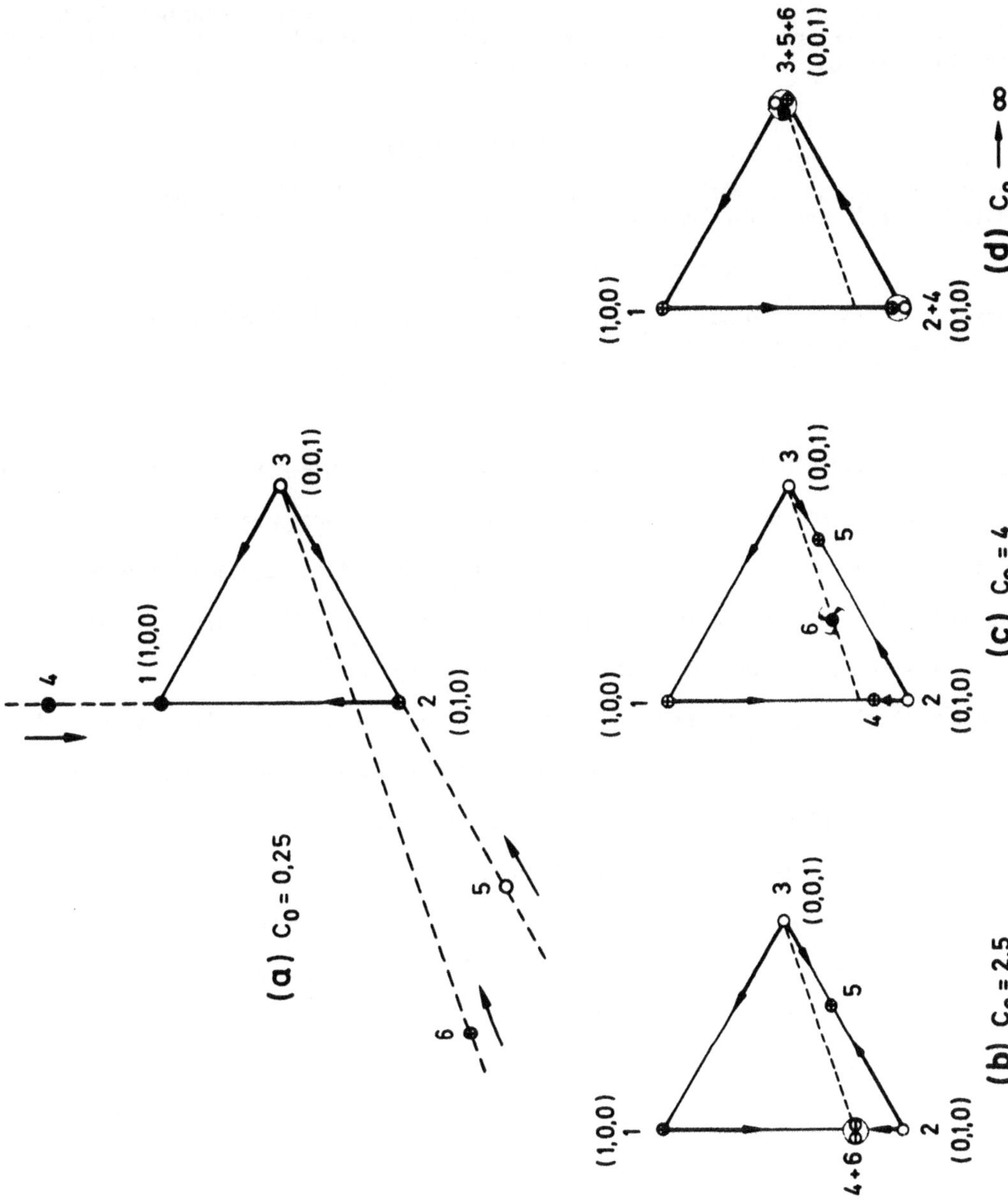
(a) $c_0 = 0{,}25$
(b) $c_0 = 2{,}5$
(c) $c_0 = 4$
(d) $c_0 \longrightarrow \infty$
1 (1,0,0)
2 (0,1,0)
3 (0,0,1)
4
5
6
4+6
2+4
3+5+6

Fig. 8.6. Three-dimensional fixed-point map for cyclically coupled chains of information carriers. The dynamics of the system are described by the differential equations (8.40). In the example shown here the rate constants have been given the values:

$$k_1=1,\ k_2=2,\ k_3=3 \text{ and } k_1'=1,\ k_2'=2,\ k_3'=3$$

The values for the parameter c_0 correspond to (a) low and (c) high concentrations, while Fig. 8.6b shows the behaviour of the system at the critial value

$$c_0 = c_{cr} = k_3(k_1'^{-1}+k_2'^{-1}) - k_1 k_1'^{-1} - k_2 k_2'^{-1}$$

Details of the fixed-point analysis may be found in Chapter 9 and Appendix A.2. (From Eigen and Schuster, 1979.) ▶

(3) CYCLICALLY-COUPLED CHAINS OF INFORMATION CARRIERS (HYPERCYCLE). The above type of coupling had the disadvantage that the growth of the first member of the catalytic chain received no assistance from the other information carriers. However, this disadvantage is eliminated if we allow the last member of the chain to influence favourably the growth of the first.

A cyclically-coupled array of self-reproducing molecular species is called a *hypercycle* (Fig. 8.3c). As we shall see, hypercycles have the exact property which we have been looking for: they are able to overcome the information threshold and thereby make possible the build-up of integrated genetic systems.

Let us first demonstrate that cyclic coupling leads to the co-operative selection of all the units involved. For simplicity's sake we consider in Fig. 8.3c an *unbranched* hypercycle. The growth functions for species I_2 to I_N are the same as those in the case of linear coupling (*cf.* Eq. 8.29). For species I_1 we have to add to the usual first-order growth term (*cf.* Eq. 8.28) a further, catalytic term. We thus obtain for the hypercyclic system in Fig. 8.3c the growth function:

$$\Gamma_i = k_i x_i + k_i' x_i x_j \qquad \begin{bmatrix} j = i-1 + N\delta_{i1} \\ i = 1,\ldots\ldots,N \end{bmatrix} \tag{8.39}$$

This growth function represents a generalization of the function (8.29) for cyclically closed chains. In the case $k_1 = 0$ the ring is broken and the functions (8.39) are identical to those for acyclic chains. By using the *Ansatz* (8.39) we

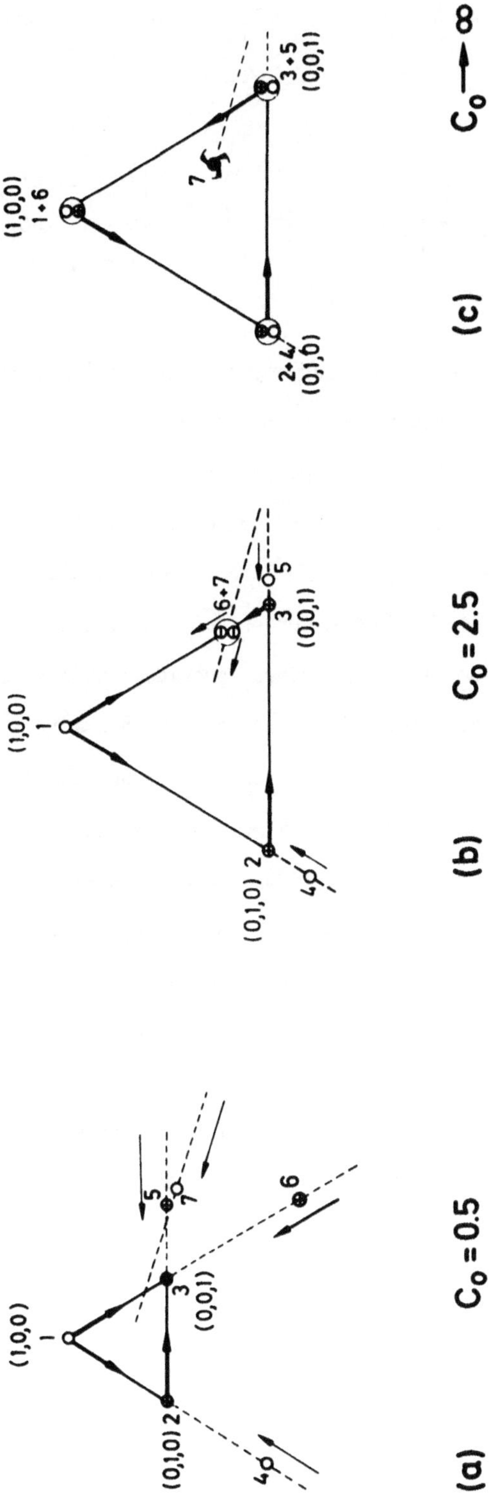
(1,0,0)
1
(0,1,0) 2
3
(0,0,1)
4
5
6
7
(a)
$C_o = 0.5$
(1,0,0)
1
(0,1,0) 2
4
6+7
3
(0,0,1)
5
(b)
$C_o = 2.5$
(1,0,0)
1+6
2+4
(0,1,0)
3+5
(0,0,1)
7
(c)
$C_o \rightarrow \infty$

obtain from Eq. (8.8) the CP selection equations for a hypercyclically coupled system of information carriers

$$\dot{x}_i = k_i x_i + k_i' x_i x_j - \frac{x_i}{c_0} \sum_l (k_l x_l + k_l' x_l x_k) \qquad \begin{bmatrix} j = i-1+N\delta_{i1} \\ k = l-1+N\delta_{l1} \\ i = 1,\ldots\ldots,N \end{bmatrix} \tag{8.40}$$

Although the dynamics of cyclic and acyclic information chains have many similarities, their long-term behaviour differs markedly. This will now be shown again by a fixed-point analysis of the three-component system.

The positions of the fixed points here too depend on the total concentration c_0. This dependence is shown in Fig. 8.6 for the three-dimensional case. Of the system's seven fixed points, three lie in the corners of the concentration simplex, while the remaining ones are for low c_0 values outside the physically meaningful region (Fig. 8.6a). As c_0 rises (Fig. 8.6b), the fixed points $\mathbf{x}^{S_4}$, $\mathbf{x}^{S_5}$, $\mathbf{x}^{S_6}$ and $\mathbf{x}^{S_7}$ move towards the simplex. For high values of c_0 (Fig. 8.6c) the fixed points $\mathbf{x}^{S_1}$, $\mathbf{x}^{S_2}$ and $\mathbf{x}^{S_3}$ lie at the corners, $\mathbf{x}^{S_4}$, $\mathbf{x}^{S_5}$ and $\mathbf{x}^{S_6}$ on the edges and $\mathbf{x}^{S_7}$ in the interior.

There is thus a fundamental difference between the long-term behaviour of the cyclic and that of the acyclic system: while the acyclic system leads with rising c_0 to selection (only component I_3 survives), the cyclic system leads to fixed relative concentrations of all of the individuals that take part in the hypercycle (I_1, I_2 and I_3 all survive).

A hypercyclically coupled system of information carriers therefore possesses the information-integrating properties which we are looking for: the system organizes itself spontaneously according to the dynamic pattern expressed in Eq. (8.40) and ends up in a state of dynamically controlled production of all of the molecular species taking part.

9. Information-Integrating Mechanisms

The fixed-point analysis has shown that the catalytic hypercycle can allow co-existence and thus make possible the progressive evolution of a whole class of non-related information carriers. Before investigating in detail the dynamic properties of hypercycles, we first note the variety of reaction networks that can be based upon hypercyclic organization patterns.

9.1. The Hypercycle - Basic Principles

For simplicity, consider hypercycles whose growth functions Γ_i are given exclusively by catalytic terms of the form

$$\Gamma_i = k_i \prod_{\alpha=1}^{n} x_\alpha^{p_{\alpha i}} \tag{9.1}$$

This excludes, for example, the system defined by the functions (8.39), since the latter contain first-order terms as well. Each hypercycle is built up from n molecular species. The quantity $p_{\alpha i}$ is an integer and gives the power to which the concentration variable x_α is raised in the ith growth function of a hypercycle. In the following treatment we shall consider only *homogeneous* growth functions, for which, by definition (*cf.* Section 8.1)

$$\sum_{\alpha=1}^{n} p_{\alpha i} = p_i \equiv p \qquad (i = 1,\ldots\ldots,n) \tag{9.2}$$

We shall also assume that the *individual* concentration terms in Γ_i are of at most first order. Thus the growth powers can only adopt values allowed by

$$p_{\alpha i} = 0 \text{ or } 1 \qquad (\alpha,\ i = 1,\ldots\ldots,n) \tag{9.3}$$

Figure 9.1 shows schematically three hypercyclic reaction networks of increasing complexity, expressed by different matrices

$$\mathbf{p} = \mathbf{p}(n,p) = (p_{\alpha i}) \tag{9.4}$$

Systems for which $p=1$ contain only linear growth terms and therefore cannot be classified as hypercycles. Conversely, hypercycles are defined by the condition

$$n \geq 2 \geq p \tag{9.5}$$

We finally introduce a cyclical symmetry into the growth function Γ_i by setting

$$\Gamma_i = k_i x_i x_j x_k \cdot \ldots \cdot x_r \qquad \begin{bmatrix} j = i - 1 + n\delta_{i1} \\ k = i - 2 + n(\delta_{i1} + \delta_{i2}) \\ \vdots \\ r = i - p + n \sum_{\mu=1}^{p-1} \delta_{i\mu} \end{bmatrix} \tag{9.6}$$

The assumption of symmetry does not restrict the general applicability of the solutions. It is however reasonable in the absence of specific information about the kinetic equations of the individual information carriers or possible differences between them.

The *Ansatz* (9.6) reduces the CP selection equations (8.8) to

$$\dot{x}_i = k_i x_i x_j \cdot \ldots \cdot x_l - \frac{x_i}{c_0} \sum_{r=1}^{n} k_r x_r x_s \cdot \ldots \cdot x_t \tag{9.7}$$

with the range of subscripts

$$\begin{aligned} j &= i - 1 + n\delta_{i1} \\ l &= i - p + 1 + n \sum_{\mu=1}^{p-1} \delta_{i\mu} \\ s &= r - 1 + n\delta_{r1} \\ t &= r - p + 1 + n \sum_{\mu=1}^{p-1} \delta_{r\mu} \\ p &\leq n \end{aligned} \tag{9.8}$$

Fig. 9.1. Hypercyclic organization of various degrees of complexity. The schemes show 6-membered hypercycles with different degrees of coupling. For each hypercycle the matrix $\mathbf{p} = (p_{\alpha i})$ is given. (From Eigen and Schuster, 1979.)

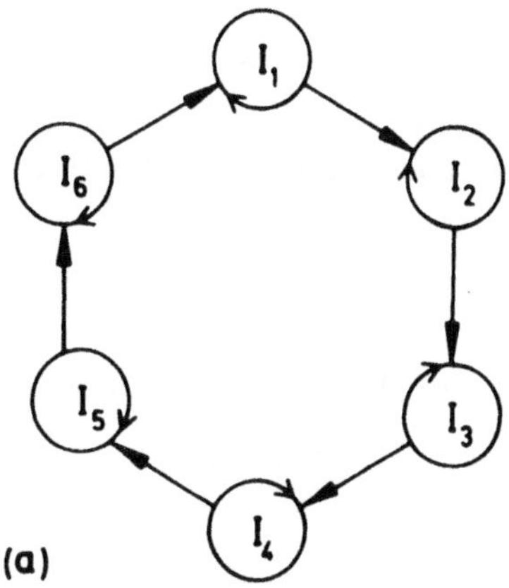

(a)

(a) Elementary hypercycle (p=2):

$$\mathbf{p} = \begin{pmatrix} 1&0&0&0&0&1 \\ 1&1&0&0&0&0 \\ 0&1&1&0&0&0 \\ 0&0&1&1&0&0 \\ 0&0&0&1&1&0 \\ 0&0&0&0&1&1 \end{pmatrix}$$

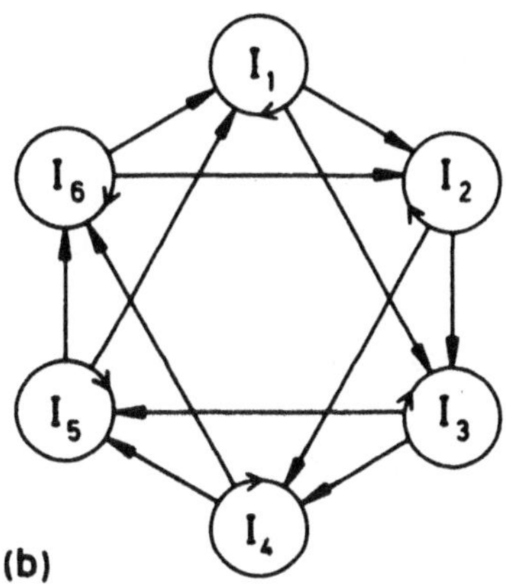

(b)

(b) Partially coupled hypercycle (p=3):

$$\mathbf{p} = \begin{pmatrix} 1&0&0&0&1&1 \\ 1&1&0&0&0&1 \\ 1&1&1&0&0&0 \\ 0&1&1&1&0&0 \\ 0&0&1&1&1&0 \\ 0&0&0&1&1&1 \end{pmatrix}$$

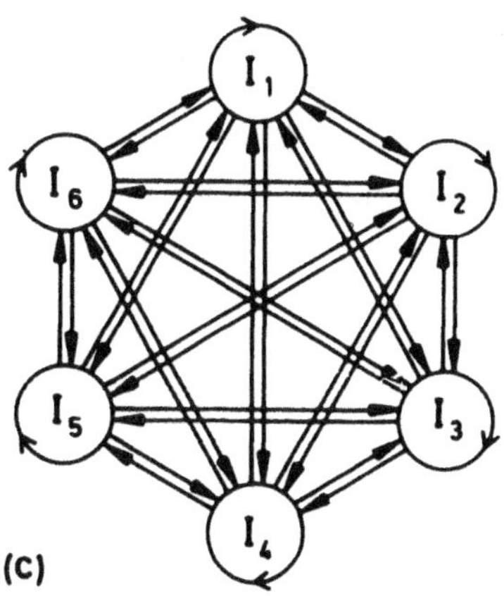

(c)

(c) Fully coupled hypercycle (p=6). All members are catalytically coupled to each other, *i.e.*, $p_{\alpha i}$=1 for all indices α and i:

$$\mathbf{p} = \begin{pmatrix} 1&1&1&1&1&1 \\ 1&1&1&1&1&1 \\ 1&1&1&1&1&1 \\ 1&1&1&1&1&1 \\ 1&1&1&1&1&1 \\ 1&1&1&1&1&1 \end{pmatrix}$$

We now proceed to a fixed-point analysis of the dynamics of hypercyclic reaction systems. To simplify the treatment, we assume that all the rate constants are equal:

$$k_1 = \ldots\ldots = k_n \equiv k \tag{9.9}$$

Calculation of the fixed points (Schuster *et al.*, 1977) then leads to the following results:

(1) There is always one fixed point ($\mathbf{x}^{s_0}$) which lies in the interior of the simplex and expresses co-existence of all members of the cycle. The condition (9.9) sets $\mathbf{x}^{s_0}$ in the middle of the simplex:

$$\mathbf{x}^{s_0} = \begin{pmatrix} \frac{c_0}{n} \\ \vdots \\ \frac{c_0}{n} \end{pmatrix} \tag{9.10}$$

(2) There are n fixed points lying at the apices of the concentration simplex.

(3) In many cases there are further one-, two-, three- and higher-dimensional sets of fixed points, which always lie on the periphery of the appropriate simplex.

We are naturally most interested in the fixed point $\mathbf{x}^{s_0}$, which lies inside the simplex and which characterizes the co-operative behaviour of the system. A linear stability analysis for this point gives a spectrum of n eigenvalues

$$\omega_\alpha^{(0)} = -\frac{1-\gamma^{-(p-1)}}{1-\gamma}\left(\frac{c_0}{n}\right)^{p-1} k \qquad (\alpha = 1,\ldots\ldots,n-1) \tag{9.11}$$

and

$$\omega_n^{(0)} = -\left(\frac{c_0}{n}\right)^{p-1} k \tag{9.12}$$

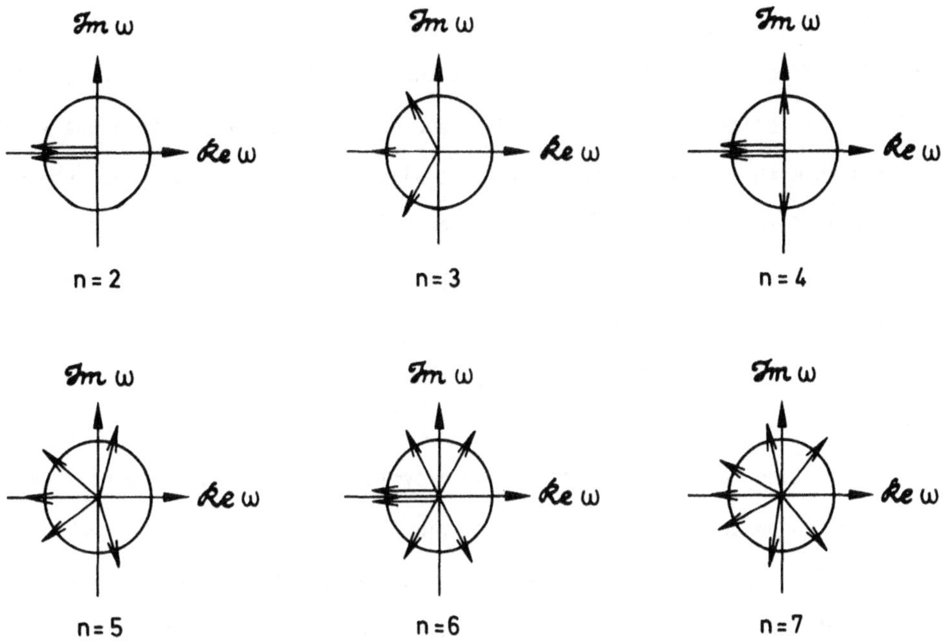

Fig. 9.2. Topological nature of the central fixed point $\mathbf{x}^{S_0}$ (Eq. 9.10) and its dependence upon the dimension of the system. Linear stability analysis about the point $\mathbf{x}^{S_0}$ gives a spectrum of complex and real eigenvalues $\omega_n^{(0)}$, which are represented as vectors in the Gaussian plane (*e.g.* for n=3 there are one real and two complex eigenvalues). The system for which results are shown is the hypercycle defined by Eq. (9.7) with degenerate rate constants k_i and degree p=2. (From Eigen and Schuster, 1979.)

with

$$\gamma = \exp\left(\frac{2\pi i}{n}\alpha\right) \tag{9.13}$$

The topological nature of the fixed point $\mathbf{x}^{S_0}$ therefore depends on the dimension n of the systems. To demonstrate this more clearly, Fig. 9.2 shows the eigenvalue spectrum $\omega_n^{(0)}$ for various dimensions (n = 2,3,.....,7) in the *Gaussian plane*. Each eigenvalue is represented here by a two-component vector

$$\boldsymbol{\omega} = \omega_{Re}\,\mathbf{e}_1 + \omega_{Im}\,\mathbf{e}_2 \tag{9.14}$$

whose first component is the real and whose second is the imaginary part. According to the classification in Table 8.1, for n=2 the central fixed point is a focus, for n=3 it is a spiral and for n=4 a centre. For higher dimensions we obtain saddle-points with spiral components in some planes. An exact analysis shows that the central fixed point $\mathbf{x}^{S_0}$ is asymptotically* stable for n=2, 3 and 4. For $n \geq 5$ it describes a stable closed orbit or limit cycle, which however always remains entirely within the simplex and never reaches its boundaries.

Let us finally discuss the probability of nucleation of hypercycles of various degrees n.

The nucleation of an elementary hypercycle (Fig. 9.1a) clearly requires bimolecular encounters between its individual components. In a fully coupled hypercycle (Fig. 9.1c), however, every component must contribute to the rate of formation of every other component. This requires either a (highly improbable) multimolecular collision or the formation of an intermediate complex composed of n components. The formation of this kind of complex will naturally be greatly restricted if the concentrations of its components are low, as must have been the case during the prebiotic phase of evolution. The nucleation of a fully coupled hypercycle therefore presupposes an extremely high and specific association constant, which could hardly arise without optimisation (see also Section 11.1).

We can thus conclude that the prebiotic hypercycle most probably had a low degree p during the phase of nucleation. It is therefore reasonable to base concrete models of prebiotic hypercycles upon the elementary case of $p=2$. However, before discussing concrete models we need to find out more about the dynamics of hypercycles.

---

*The concept "asymptotically stable" is defined in Appendix A.2.

9.2. Dynamics of Elementary Hypercycles

A complete, analytical description of *elementary hypercycles* ($p=2$; *cf.* Fig. 9.1a) can easily be performed for systems of dimensions $n \leq 4$. For higher-dimensional hypercycles numerical methods must be used. To preserve some degree of uniformity we shall first solve Eqs. (9.7) numerically for all the dimensions considered heretofore. An analytical treatment of hypercycles with dimensions $n \leq 4$ may be found in the paper of M. Eigen and P. Schuster (1979).

From Eq. (9.7) the dynamics of elementary hypercycles of dimension n are described by the following set of differential equations:

$$\dot{x}_i = \Lambda_i(x_1, \ldots\ldots, x_n) \qquad (i = 1, \ldots\ldots, n) \tag{9.15}$$

with the functions

$$\Lambda_i = k_i x_i x_j - \frac{x_i}{c_0} \sum_{r=1}^{n} k_r x_r x_s \qquad \begin{bmatrix} j = i-1+n\delta_{i1} \\ s = r-1+n\delta_{r1} \end{bmatrix} \tag{9.16}$$

In the fixed-point analysis discussed above we assumed equal rate constants ($k_1 = \ldots\ldots = k_n \equiv k$). However, it should here again be emphasised that the general form of the solutions is more or less independent of the relative magnitudes of the rate constants (Eigen and Schuster, 1979). For the sake of simplicity we shall retain the assumption of degenerate rate constants in the following analysis.

We shall investigate two aspects of the set of differential equations (9.15): both the time-dependent solutions

$$x_i = x_i(t) \qquad (i = 1, \ldots\ldots, n) \tag{9.17}$$

and the trajectories in n-dimensional concentration space

$$x_i = x_i(x_1, \ldots\ldots, x_{i-1}, x_{i+1}, \ldots\ldots, x_n) \qquad (i = 1, \ldots\ldots, n) \tag{9.18}$$

will be determined. The latter presentation serves to eliminate the explicit time-dependence from Eqs. (9.17).

The differential equations for the trajectories result from Eqs. (9.15) by simple division:

$$\frac{\dot{x}_i}{\dot{x}_1} = \frac{dx_i}{dx_1} = \frac{\Lambda_i}{\Lambda_1} = f_i(x_1,\ldots\ldots,x_n) \qquad (i = 2,\ldots\ldots,n) \tag{9.19}$$

This is a set of (n-1) differential equations, whose integration gives (n-1) trajectories:

$$\begin{aligned} x_2 &= x_2(x_1,x_3,\ldots\ldots,x_n) \\ x_3 &= x_3(x_1,x_2,x_4,\ldots,x_n) \\ &\vdots \\ x_n &= x_n(x_1,\ldots\ldots\ldots,x_{n-1}) \end{aligned} \tag{9.20}$$

As long as the trajectories run completely within the appropriate concentration simplex, the result is the co-existence of all the partners in the hypercycle.

Let us examine some elementary hypercycles using the criterion of co-existence. Since trajectories can only be satisfactorily shown in two dimensions, we shall show their projections onto the plane containing x_1 and one of the other co-ordinates x_k.

The topological nature of the central fixed point has already been investigated for elementary hypercycles of various dimensionalities in the context of Fig. 9.2. We shall now complete this investigation by a more detailed analysis of the dynamics involved. Values of the parameters used are given in the captions of Figs. 9.3 to 9.8.

(1) ELEMENTARY HYPERCYCLES WITH n=2. The concentration variables x_i tend toward a constant equilibrium value (Fig. 9.3a). The trajectory in the (x_1, x_2)-plane runs directly into a focus (Fig. 9.3b).

(2) ELEMENTARY HYPERCYCLES WITH n=3. The concentration variables x_i show damped oscillations (Fig. 9.4a). The trajectories run along spirals into the central fixed

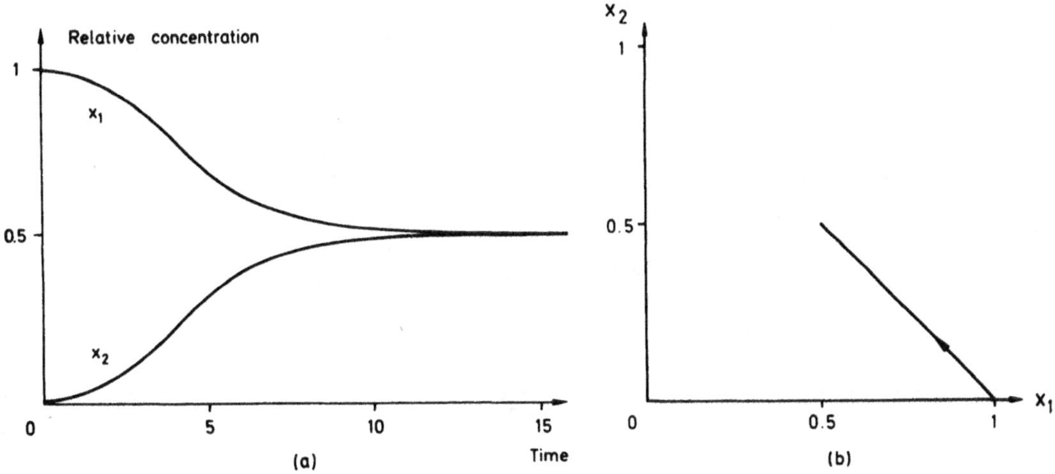

Fig. 9.3. Dynamics of an elementary hypercycle of dimension n=2. Initial conditions: $\mathbf{x}(0) = (0.99, 0.01)$, rate constants: $\mathbf{k} = (1,1)$. (a) Time-dependent solutions, (b) trajectory in the (x_1,x_2)-plane.

point (Fig. 9.4b). The fixed point is approached fairly quickly, as is implied by the strong damping of the oscillations of the time-dependent solution $\mathbf{x}(t)$.

(3) ELEMENTARY HYPERCYCLES WITH n=4. The concentration variables x_i undergo stationary oscillations (Fig. 9.5a). As in the three-dimensional case, the trajectories run along a spiral towards the central fixed point (Fig. 9.5b). However, in this case the centripetal tendency is much weaker.

(4) ELEMENTARY HYPERCYCLES WITH n=5. The concentration variables x_i perform stationary oscillations (Fig. 9.6a). The central fixed point is a saddle and is therefore unstable. Since there are no sinks on the boundary of the concentration simplex, the trajectories must be stable closed orbits which run entirely in the interior of the concentration simplex. Even if the existence of an attractor of this kind for higher-dimensional hypercycles ($n \geq 5$) has not yet been analytically proven, the numerical results in Figs. 9.6b and 9.7b still support our notion of the existence of such an attractor. It is clear that the dynamics of the system tend after a sufficiently long time to the same limit cycle (compare Figs. 9.6c and 9.6d), which

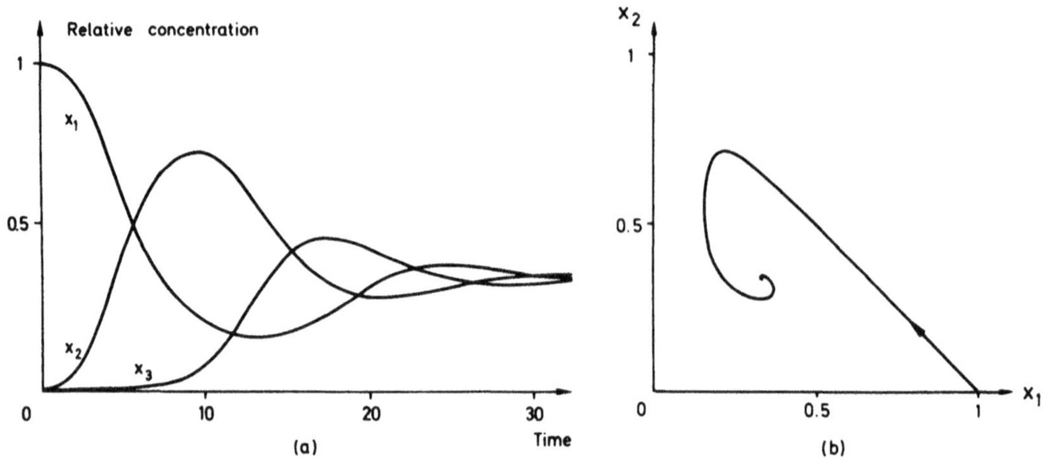

Fig. 9.4. Dynamics of an elementary hypercycle of dimension n=3. Initial conditions: $\mathbf{x}(0) = (0.98, 0.01, 0.01)$, rate constants: $\mathbf{k} = (1,1,1)$. (a) Time-dependent solutions, (b) trajectory in the (x_1,x_2)-plane.

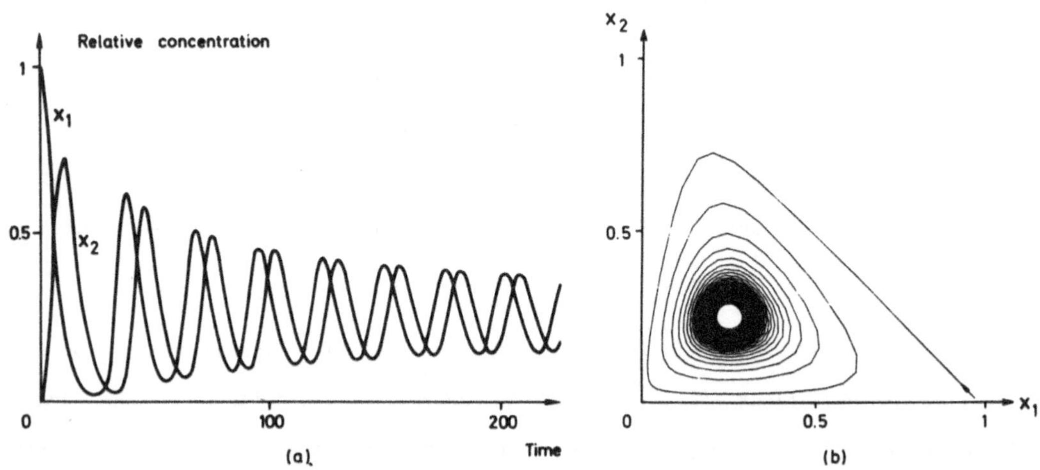

Fig. 9.5. Dynamics of an elementary hypercycle of dimension n=4. Initial conditions: $\mathbf{x}(0) = (0.97, 0.01, 0.01, 0.01)$, rate constants: $\mathbf{k} = (1,1,1,1)$. (a) Time-dependent solutions, shown for x_1 and x_2; (b) trajectory in the (x_1, x_2)-plane.

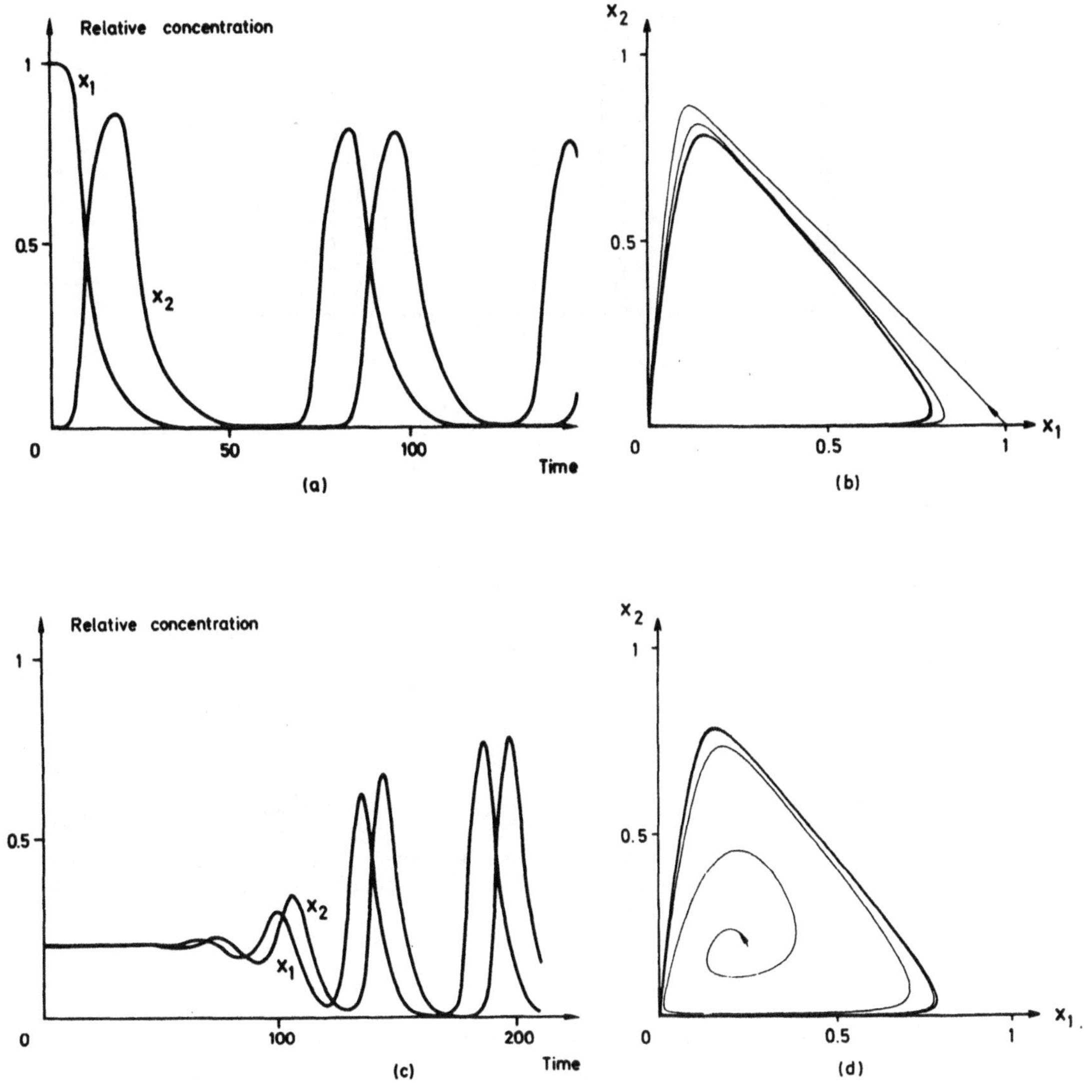

Fig. 9.6. Dynamics of an elementary hypercycle of dimension n=5. Initial conditions: $\mathbf{x}(0) = (0.9996, 0.0001,\ldots\ldots, 0.0001)$, rate constants: $\mathbf{k} = (1,1,1,1,1)$. (a) Time-dependent solutions, shown for x_1 and x_2; (b) trajectory of these in the (x_1,x_2)-plane, (c) time-dependent solutions for x_1 and x_2 under the altered initial conditions $\mathbf{x}(0) = (0.2004, 0.1999,\ldots\ldots, 0.1999)$, (d) trajectory of these in the (x_1, x_2)-plane.

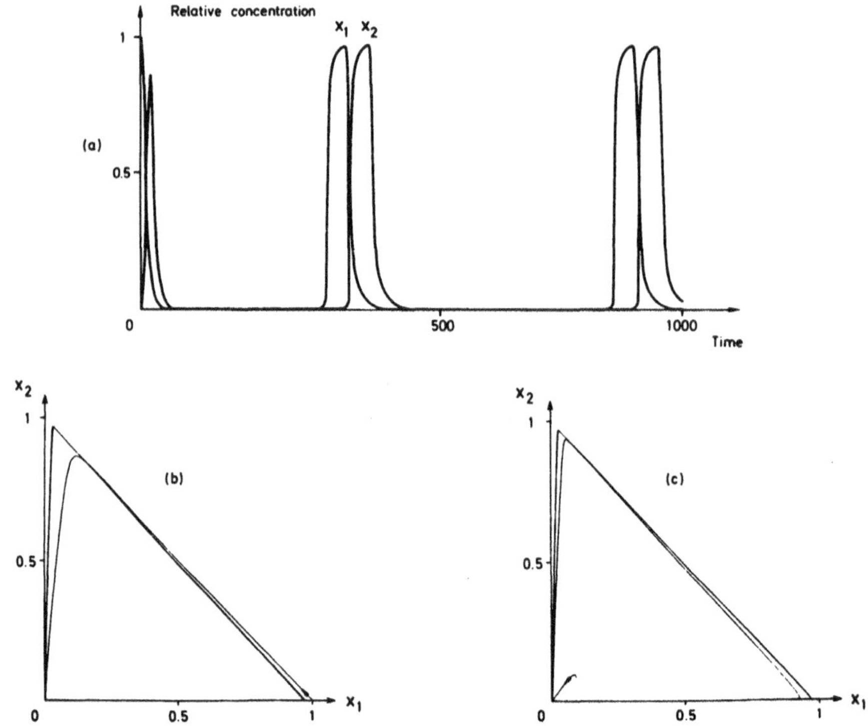

Fig. 9.7. Dynamics of an elementary hypercycle of dimension n=12. Initial conditions: $\mathbf{x}(0)$ = (0.9989, 0.0001,....., 0.0001), rate constants: $\mathbf{k}$ = (1,.....,1). (a) Time-dependent solutions; only x_1 and x_2 are shown, (b) trajectory in the (x_1,x_2)-plane, (c) trajectory in the (x_1,x_2)-plane under altered initial conditions $\mathbf{x}(0)$ = (0.0848, 0.0832,....., 0.0832).

is independent of the starting conditions. For example, if we have a uniform concentration distribution and subject it at time t=0 to a small perturbation, then the oscillations build up spontaneously (Fig. 9.6b).

(5) ELEMENTARY HYPERCYCLES WITH n=12. Hypercycles with a dimension $n \geq 5$ all show the same qualitative dynamical picture. The conclusions reached for the 5-dimensional hypercycle retain their validity. A comparison between Figs. 9.6a and 9.7a shows, however, that with increasing n the oscillations in individual concentration variables approach the form of square pulses. The geometry of the limit cycles changes concomitantly (Fig. 9.7b). The system continues to approach the same limit cycle irrespective of its starting-point.

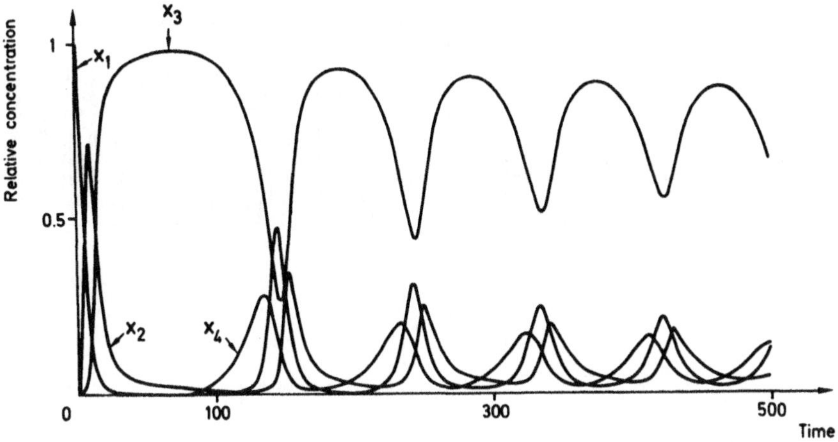

Fig. 9.8. Time-dependent solutions for a four-dimensional hypercycle with an asymmetric distribution of rate constants. Initial conditions: $\mathbf{x}(0) = (0.97, 0.01, 0.01, 0.01)$, rate constants: $\mathbf{k} = (1,1,1, 0.1)$.

Let us summarize our results. All elementary hypercycles show the same dynamic behaviour, irrespective of their dimension: *each hypercycle possesses exactly one attractor, which lies inside the concentration simplex and demonstrates the co-existence of all the components of the hypercycle.* If the dimension number is low (n=2,3), the concentration distribution at equilibrium is constant. For n=3 the equilibrium is reached by way of damped oscillations. For $n \geq 4$, on the other hand, stationary oscillations occur. However even for $n \geq 4$ all components show co-existence, although in contrast to the lower-dimensional cases (n=2,3) the balance of this co-existence is time-dependent in a way determined by the overall dynamics of the system.

For asymmetric distributions of the rate constants $\mathbf{k}$ we obtain a similar result. As an example, the stationary oscillations of a four-dimensional hypercycle are shown in Fig. 9.8. The rate constant for formation of the fourth component has been set at one-tenth of the rate constants of the other components. In selection equilibrium the concentration of the component from which the fourth is formed (*i.e.*, of the third component) must therefore be greater on average than that of the other

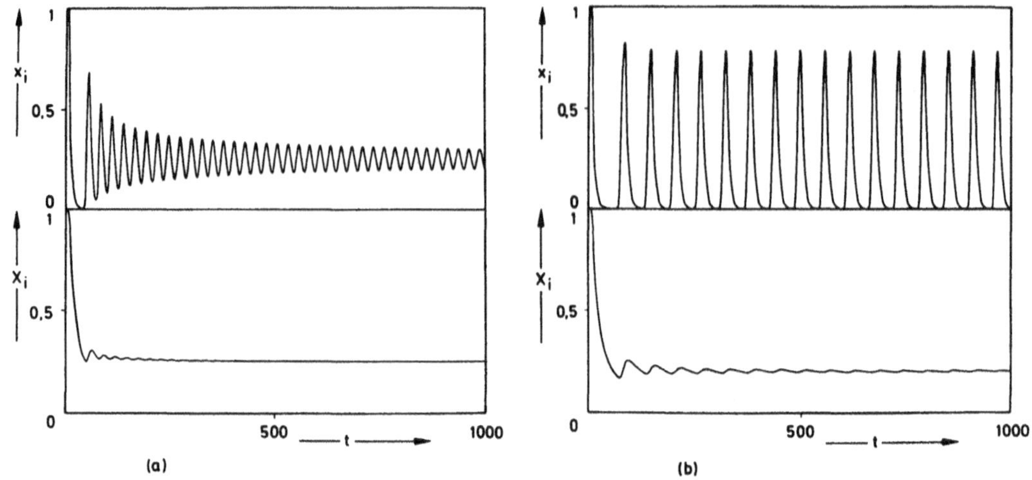

Fig. 9.9. Attainment of internal equilibrium in elementary hypercycles. The concentration variables $x_i(t)$ and their time-averaged values $X_i(t)$ are shown. (a) Hypercycle of dimension n=4. Initial conditions: $\mathbf{x}(0) = (0.97, 0.01, 0.01, 0.01)$, rate constants: $\mathbf{k} = (1,1,1,1)$. (b) Hypercycle of dimension n=5. Initial conditions: $\mathbf{x}(0) = (0.9996, 0.0001,\ldots\ldots, 0.0001)$, rate constants: $\mathbf{k} = (1,1,1,1)$. (From Eigen and Schuster, 1979.)

partners. A similar result has been obtained for the process of equilibration in a self-replicating (±) ensemble (Section 5.6).

We shall finally look at the question of how rapidly equilibrium is reached between the individual components in a growing hypercycle. For low-dimensional hypercycles (n=2,3) the rate of attainment of equilibrium can be seen directly in Figs. 9.3 and 9.4. In oscillating systems of dimension $n \geq 4$, by contrast, we must first obtain a time-averaged value of each concentration variable x_i using the relation

$$X_i(t) = \lim_{t\to\infty} \frac{1}{t} \int_0^t x_i(\tau)d\tau \tag{9.21}$$

in order to assess the extent of equilibration. In the case of degenerate rate constants the solutions for all the components of the cycle are the same, so it will suffice to show the time-average of any one of them (Fig. 9.9). As is seen, even in cases where stationary oscillations occur, equilibrium is reached after a few generation times. The time-averaged x_i-variables coincide with the co-ordinates of the central fixed point (Eq. 9.10). Thus, for example, in the cases shown in Fig. 9.9, after only a few oscillations the time-averaged value $X_i(t)$ has already reached the value 0.25 (for n=4 and c_0=1) or 0.20 (for n=5 and c_0=1). We shall later make frequent use of the rapid attainment of equilibrium in hypercycles in order to reduce the number of independent variables.

9.3. Selection Properties of Hypercycles

Now that we have discussed in some detail the dynamics of single hypercycles, we can look at some selection models for mixed populations of different hypercycles. To begin with we shall analyse the kinetics of selection of independently-growing hypercycles (Fig. 9.10a). We shall then look for the conditions under which coupled growth between hypercycles might occur (Fig. 9.10c). One primitive form of this kind of reaction network is the formation of parasites within a hypercycle (Fig. 9.10b).

(1) INDEPENDENTLY-GROWING HYPERCYCLES. Consider a population of m independently-growing hypercycles of dimension n (Fig. 9.10a). In a system free of constraint the growth equations for the kth hypercycle are

$$\dot{x}_{i\alpha} = k_{i\alpha} x_{i\alpha} x_{j\alpha} - \varphi x_{i\alpha} \qquad \begin{bmatrix} i = 1,\ldots\ldots,n \\ j = i - 1 + n\delta_{i1} \\ \alpha = 1,\ldots\ldots,m \end{bmatrix} \tag{9.22}$$

Here the quantities x, k and φ have again their usual meaning: $x_{i\alpha}$ is the concentration of the ith component of the kth hypercycle and $k_{i\alpha}$ is the rate constant for its formation. φ is a control variable that can be used to impose particular constraints (*e.g.* CP conditions) on the system.

We now introduce the CP constraint

$$\sum_\alpha \sum_i x_{i\alpha} = c_0 = \text{const.} \tag{9.23}$$

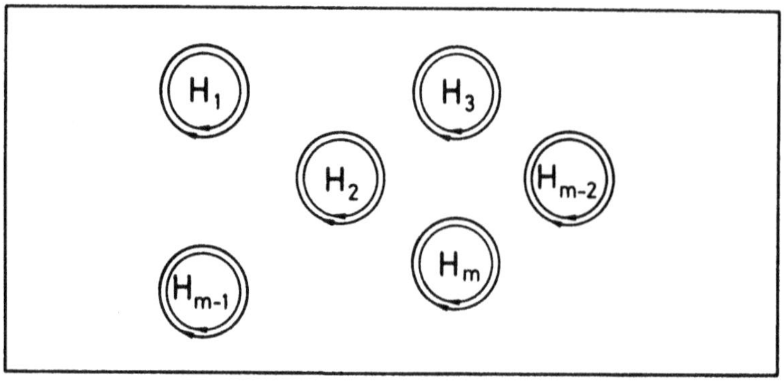

(a) Independently-growing hypercycles

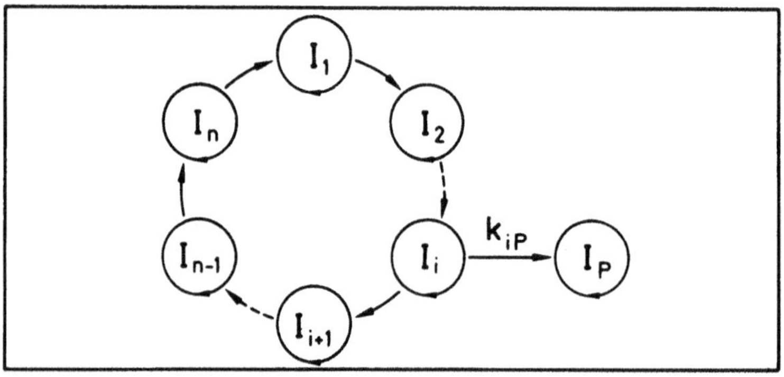

(b) Hypercycle with a parasitic component

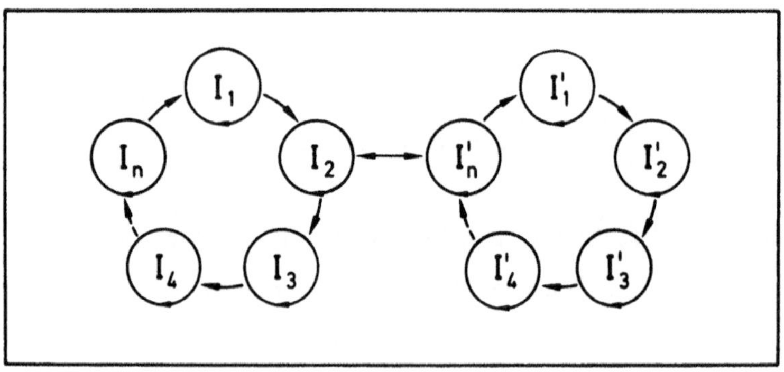

(c) Coupled hypercycles

Fig. 9.10. Networks of hypercycles.

and obtain from Eq. (9.22)

$$\sum_a\sum_i \dot{x}_{ia} = 0 = \sum_a\sum_i k_{ia}x_{ia}x_{ja} - \varphi\sum_a\sum_i x_{ia} \tag{9.24}$$

In order for the system to fulfil CP conditions, φ must satisfy the relation

$$\varphi = \frac{1}{c_0}\sum_a\sum_i k_{ia}x_{ia}x_{ja} \tag{9.25}$$

The CP selection equations are then (*cf.* Eq. 8.40):

$$\dot{x}_{ia} = k_{ia}x_{ia}x_{ja} - \frac{x_{ia}}{c_0}\sum_a\sum_i k_{ia}x_{ia}x_{ja} \tag{9.26}$$

The selection kinetics defined by this set of equations are shown in Fig. 9.11 for competition between three hypercycles. For comparison, Fig. 9.12 shows the corresponding selection kinetics for a non-coupled system of information carriers. Their selection equations are obtained from Eq. (9.26) for the special case where n=1. Thus

$$\dot{x}_i = k_ix_i - \frac{x_i}{c_0}\sum_j k_jx_j \tag{9.27}$$

The structure of this equation is naturally identical with that of Eq. (5.74), which describes the selective growth of a quasi-species under CP conditions.

In spite of identical initial conditions and identical parameter values, the selection kinetics differ greatly between the two systems. Selection between hypercycles is much sharper than selection between non-coupled information carriers. In the latter, mutants with a selective advantage can always assert themselves, whereas hypercyclic selection necessarily leads to a "once-for-all-time" decision (*cf.* Section 9.4).

These remarkable differences in selection characteristics can be understood in terms of the growth laws underlying the corresponding selection equations. For this

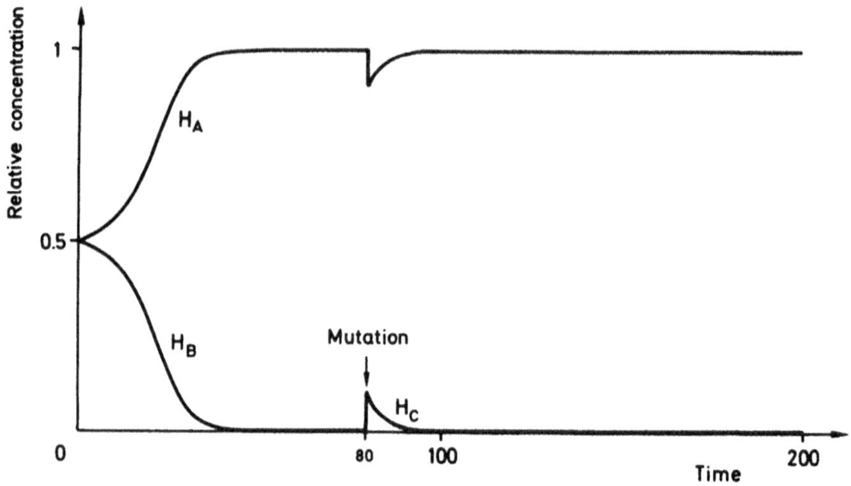

Fig. 9.11. Competition between three elementary hypercycles under CP conditions. For each hypercycle H_A, H_B, H_C the total concentration of all its components is shown as a function of time. *E.g.* for H_A this is $c_A(t) = \sum_i x_{Ai}(t)$.

At time t=0 only H_A and H_B are present and their concentrations are equal. Initial conditions: $\mathbf{x}_A(0) = \mathbf{x}_B(0) = (0.125,.......,0.125)$; rate constants: $\mathbf{k}_A = (1,....., 1)$, $\mathbf{k}_B = (0.9,....., 0.9)$. The hypercycle H_A wins, on account of its higher growth parameter.

At time t=80 there is a mutation leading to the nucleation of a new hypercycle H_C. This mutant starts out with the initial conditions $\mathbf{x}_C(0) = (0.025,....., 0.025)$ but it possesses higher rate constants $\mathbf{k}_C = (1.1,....., 1.1)$. Although the mutant H_C has a selective advantage over H_A, it cannot - even in a deterministic model - assert itself. Hypercyclic selection always leads to a "once-and-for-all" decision. (After Eigen and Winkler-Oswatitsch, 1975.)

purpose let us first consider the growth law of an elementary hypercycle, in the absence of control terms ($\varphi \equiv 0$). The growth equations are

$$\dot{x}_i = k_i x_i x_j \qquad (i,j = 1,.....,n) \tag{9.28}$$

As soon as this hypercycle has reached equilibrium between its components, its growth as a whole can be described by a single differential equation. To derive this equation we proceed from the relationship

$$\frac{d}{dt}\left(\frac{x_i}{x_j}\right) = \frac{x_i \dot{x}_j - x_j \dot{x}_i}{x_j^2} \tag{9.29}$$

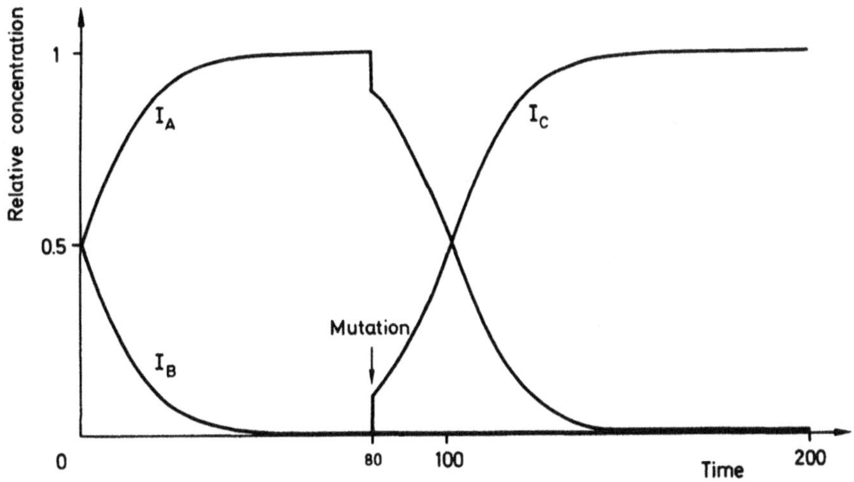

Fig. 9.12. Competition between three non-coupled information units under CP conditions.

At time t=0 only information units I_A and I_B are present. Initial conditions: $x_A(0) = x_B(0) = 0.5$; rate parameters: k_A=1.0, k_B=0.9. The information unit I_A wins owing to its superior selection properties.

At time t=80 there is a mutation leading to a new species I_C. The mutant starts with the initial value $x_C(80) = 0.1$, but it has a growth constant k_C=1.1 higher than that of I_A. It is characteristic of selection in the Darwinian sense that, in the deterministic model, a better mutant can always assert itself. (After Eigen and Winkler-Oswatitsch, 1975.)

In selection equilibrium we have

$$\frac{d}{dt}\left(\frac{x_i}{x_j}\right) = 0 \tag{9.30}$$

and this applies for all components of the hypercycle (taking averages if $n \geq 4$). It follows directly that

$$x_i \dot{x}_j = x_j \dot{x}_i \tag{9.31}$$

and after elementary rearrangement and substitution into Eq. (9.28) that

$$\frac{x_j^2}{\dot{x}_j} = \frac{1}{k_i} \tag{9.32}$$

Furthermore, in equilibrium we have

$$x_j = \frac{1}{a_j} c \tag{9.33}$$

where the a_j are constants which satisfy the relation

$$\sum_j \frac{1}{a_j} = 1 \tag{9.34}$$

Inserting Eq. (9.33) into Eq. (9.32) yields

$$\frac{1}{a_j} \frac{c^2}{\dot{c}} = \frac{1}{k_i} \tag{9.35}$$

By summarizing over all indices j and bearing in mind that $j=i-1+n\delta_{i1}$, we obtain a differential equation which describes the time-dependence of the total concentration $c=c(t)$:

$$\dot{c} = \frac{1}{\sum_i 1/k_i} c^2 = \tilde{k}c^2 \tag{9.36}$$

where $\tilde{k}$ is defined by

$$\tilde{k} = \frac{1}{\sum_i 1/k_i} \tag{9.37}$$

Since second order terms appear in the growth equations for a hypercyclically organized system, the growth is *hyperbolic*.

From the definition of $\tilde{k}$, the rate constant for growth of the complete hypercycle is of the same order of magnitude as that of the slowest individual step. Integration of Eq. (9.36) yields the time-dependent solution for c:

$$c(t) = \frac{1}{1/c_0 - kt} \tag{9.38}$$

This function possesses a singularity at time $t=t_{crit}$ given by

$$t_{crit} = \frac{1}{c_0\tilde{k}} \tag{9.39}$$

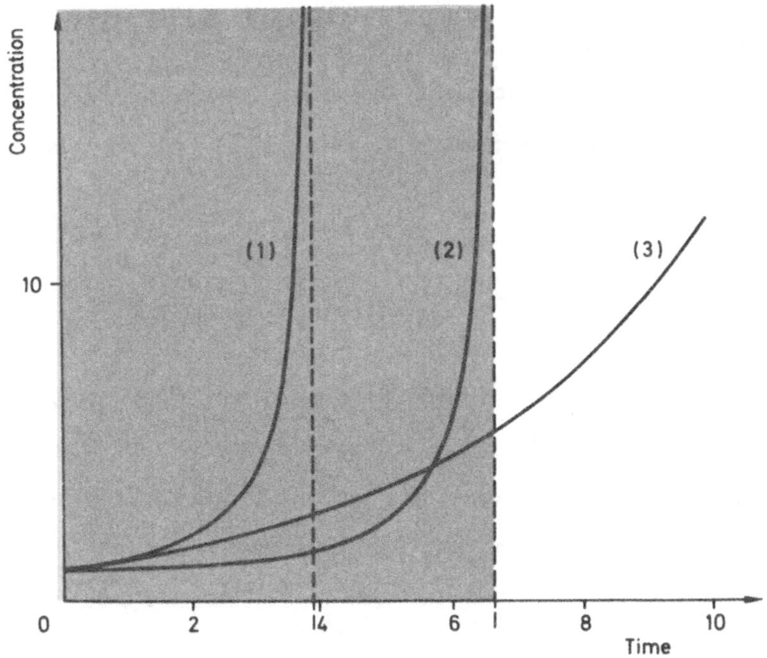

Fig. 9.13. Autocatalytic growth in a system of coupled and non-coupled information carriers. (1) Hyperbolic growth as given in Eq. 9.36, starting with a completely equilibrated elementary hypercycle (n=4, p=2). (2) Hyperbolic growth as given in Eqs. 9.28. This growth law, when applied to elementary hypercycles starting far from equilibrium, moves towards case (1) after a certain lag time for equilibration. (3) Exponential growth as given in Eq. (9.40). This is the simplest form of autocatalytic growth and occurs in systems of non-coupled information carriers.

Systems (1) and (3) start under the same initial conditions: $k = \tilde{k} = 1/4$ and $c(t=0)=1$. System (2) is not in internal equilibrium, but starts with the following initial values: $k_1=k_2=k_3=k_4=1$ and $\mathbf{x}(0) = (0.97, 0.01, 0.01, 0.01)$. At t = 3.95 and t = 6.33 singularities appear. The figure shows the time course of the total concentration, given for all systems by the sums of their components' concentrations.

At this point the model breaks down, for the population $c(t_{crit})$ tends to infinity. If the system is not in equilibrium at the beginning (t=0), the singularity occurs somewhat later.

The growth curves for an equilibrated and a non-equilibrated hypercycle are compared in Fig. 9.13. A close correspondence is seen between the two curves. In particular, both systems show the singularity at approximately the same time. The assumption that a hypercycle is in internal equilibrium from the start thus makes no qualitative difference to its overall growth properties.

We now examine the significance of the square term in Eq. (9.36) by comparing it with the corresponding first-order equation

$$\dot{c} = kc \tag{9.40}$$

which describes the growth of a system of non-coupled information carriers.

It follows from Eq. (9.40) that the self-reproduction of a non-coupled system appears as *exponential* growth. In contrast to hyperbolic growth, the population becomes infinite only after an infinitely long period. It is therefore plausible that in restricted systems hyperbolic growth must lead to sharper selection than does exponential growth.

(2) HYPERCYCLES WITH A PARASITIC COMPONENT. The schematic hypercycle of Fig. 8.3c represents the simplest form of cyclic catalysis, in which each individual component is important for the function of the whole system. However, branching can also occur in hypercycles. This takes place at points where external information units, *i.e.*, ones which were not originally part of the hypercycle, become catalytically coupled to the "basic" cycle (Fig. 9.10b). Such external components we shall denote as parasites. Parasitic coupling of arbitrary complexity is in principle conceivable. We shall now examine the case in which, for simplicity, just one parasitic component is present. The problem will be treated both analytically and numerically.

The analytical consideration will be restricted to a fixed-point analysis. To reduce the number of parameters we shall assume that the hypercycle is in equilibrium; this assumption will be dropped in the numerical treatment.

The dynamics of the equilibrated system with parasitic coupling are described by two differential equations. For the basic cycle we have the equation

$$\dot{c}_H = \tilde{k}c_H^2 - \varphi c_H \qquad (c_H = \text{concentration of the basic hypercycle}) \tag{9.41}$$

where c_H is the concentration of the basic hypercycle and $\tilde{k}$ a rate constant which is defined by Eq. (9.37). For the parasite coupled to the ith component of the basic cycle (Fig. 9.10b) we have

$$\dot{c}_P = k_{iP} x_i c_P - \varphi c_P \tag{9.42}$$

in which we have assumed that the parasite is self-reproducing. Furthermore, let us introduce a new constant k, given by

$$k = k_{iP} \frac{x_i}{c_H} \tag{9.43}$$

Equation (9.42) then becomes

$$\dot{c}_P = k c_H c_P - \varphi c_P \tag{9.44}$$

If the population of the entire system is kept constant, then

$$c_H + c_P = c_0 \tag{9.45}$$

This condition is fulfilled if the control parameter φ satisfies the condition

$$\varphi = \frac{\tilde{k} c_H^2 + k c_H k_P}{c_0} \tag{9.46}$$

Substitution of Eq. (9.46) into Eqs. (9.41) and (9.42) gives

$$\dot{c}_H = \tilde{k} c_H^2 - \frac{c_H}{c_0} (\tilde{k} c_H^2 + k c_H c_P) \tag{9.47}$$

and

$$\dot{c}_P = k c_H c_P - \frac{c_P}{c_0} (\tilde{k} c_H^2 + k c_H c_P) \tag{9.48}$$

These are the differential equations describing the selection kinetics of hypercycles with one parasite under CP conditions.

First, we can make some generalizations about the system's long-term behaviour. There are two fixed points, which lie at the apices of the S_2 simplex covering the c_H- and c_P-variables, *viz.*,

$$\mathbf{c}^{s_1} = (c_H, c_P)^{s_1} = (c_0, 0)^{s_1} \tag{9.49}$$

and

$$\mathbf{c}^{s_2} = (c_H, c_P)^{s_2} = (0, c_0)^{s_2} \tag{9.50}$$

In the special case in which $k=\tilde{k}$, there is another set of fixed points which passes through the points

$$\mathbf{c}^{s} = (c_H, c_P)^{s} = (c_H, c_0 - c_H)^{s} \tag{9.51}$$

Thus under selection pressure a hypercycle and its parasite can co-exist as long as $k=\tilde{k}$. This is none the less a special case. In real systems, we have in general $k \neq \tilde{k}$, and the hypercycle will compete with its parasites.

The fixed points defined by Eqs. (9.49) and (9.50) correspond to two eigenvalues, as follows. To $\mathbf{c}^{S_1}$ belongs

$$\omega^{(1)} = (k - \tilde{k})c_0 \tag{9.52}$$

and to $\mathbf{c}^{S_2}$ belongs

$$\omega^{(2)} = 0 \tag{9.53}$$

As we have seen, the eigenvalues give us information about the stability of the fixed points. Thus the fixed point $\mathbf{c}^{S_1}$ is stable if $\tilde{k} > k$. However the linear stability analysis is insufficient for the fixed point $\mathbf{c}^{S_2}$. Here we must introduce higher-order terms into our analysis. For the point $c_P = c_0 - \delta c_P$ we obtain from Eq. (9.48)

$$\dot{c}_P = \frac{k-\tilde{k}}{c_0} (\delta c_P)^2 (c_0 - \delta c_P) \tag{9.54}$$

It follows from this that

$$\dot{c}_P > 0 \qquad (\text{for } k > \tilde{k}) \tag{9.55}$$

and

$$\dot{c}_P < 0 \qquad (\text{for } k < \tilde{k}) \tag{9.56}$$

In the first case the fixed point $\mathbf{c}^{S_2}$ is stable and in the second it is unstable. Figure 9.14 shows three typical ways in which a hypercycle can react to parasitic coupling. Co-existence is only possible under the improbable boundary condition $k=\tilde{k}$, *i.e.*, where the values of the two rate parameters happen to coincide. In this case co-existence is possible for all arbitrary values of the variables c_H and c_P. In reality it is scarcely to be expected that k and $\tilde{k}$ have exactly the same value, especially as each of them is in turn dependent upon a number of other parameters. The case $k=\tilde{k}$ could at most be realized in systems in which the parasite I_P and its master sequence I_i make use of the same *promoter*, which for example might be localized in E_{i-1}, the translation product of I_{i-1} (*cf.* Fig. 9.16). Such a functional organization will automatically lead to a *gene-operon* structure of the code system. Parasitic branches could bring about a selective advantage in so far as functions of general utility (control factors, metabolic enzymes, components of the translation apparatus *etc.*) can evolve. However, only in a compartmented system can a parasite with advantageous translation products become optimized (Section 11.2).

Fig. 9.14. Selection dynamics of an elementary hypercycle (n=4) with a self-reproducing parasite. The figure shows the time course of the total concentration $\mathbf{c}_H$ of all the components of the basic hypercycle and of the concentration c_P of the parasite. For simplicity it was assumed that the hypercycle is already in equilibrium at time t=0. Initial conditions: $\mathbf{c}_H = (0.2,\ldots\ldots, 0.2)$, $c_P = 0.2$.

(a) If $\tilde{k} > k$ the hypercycle outgrows the parasite and the result is selection of the hypercycle. (b) In the borderline case where $\tilde{k}=k$ the hypercycle and the parasite can co-exist. (c) If $\tilde{k} < k$ the parasite is selected along with component I_i of the original hypercycle. ▶

So in real hypercycles with parasites one of the two will be selected: if $\tilde{k} > k$, the hypercycle wins and the parasite dies out, and if $\tilde{k} < k$ the opposite occurs.

(3) NETWORKS OF HYPERCYCLES. We consider a population of hypercycles and assume that the hypercycles possess catalytic factors by means of which they promote each other's growth (Fig. 9.10c). The growth-promoting factor can be a component of a neighbouring hypercycle, or it can be any substance which is produced by one of the hypercycles present. Such coupling can occur in a multitude of ways, leading to networks of hypercycles (Fig. 9.10c).

In the treatment to follow we shall look at the simplest case of a hierarchically organized population of hypercycles, *viz.*, coupling between two elementary hypercycles H_A and H_B. To simplify the discussion we shall assume that each hypercycle is in internal equilibrium. The growth equations are then

$$\dot{c}_A = \tilde{k}_A c_A^2 + k_A c_A^2 c_B^n - \varphi c_A \tag{9.57}$$

for hypercycle H_A and

$$\dot{c}_B = \tilde{k}_B c_B^2 + k_B c_B^2 c_A^m - \varphi c_B \tag{9.58}$$

for hypercycle H_B.

In this set of differential equations catalytic terms of the (n+2)th and (m+2)th order with respect to the concentrations c_A and c_B appear. The system of coupled hypercycles shall be subject to CP constraint, that is,

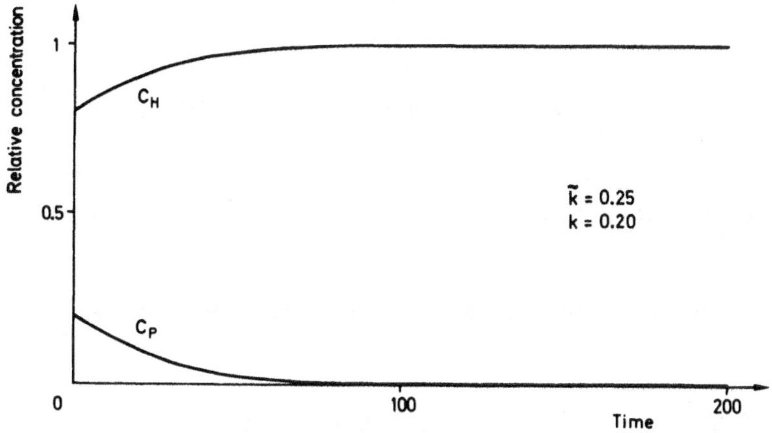
Relative concentration
1
0.5
0
100
200
Time
C_H
C_P
$\tilde{k} = 0.25$
$k = 0.20$

(a)

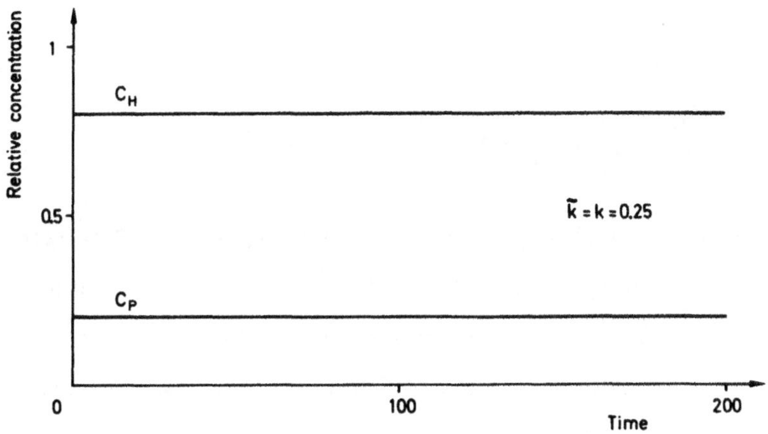
Relative concentration
1
0.5
0
100
200
Time
C_H
C_P
$\tilde{k} = k = 0.25$

(b)

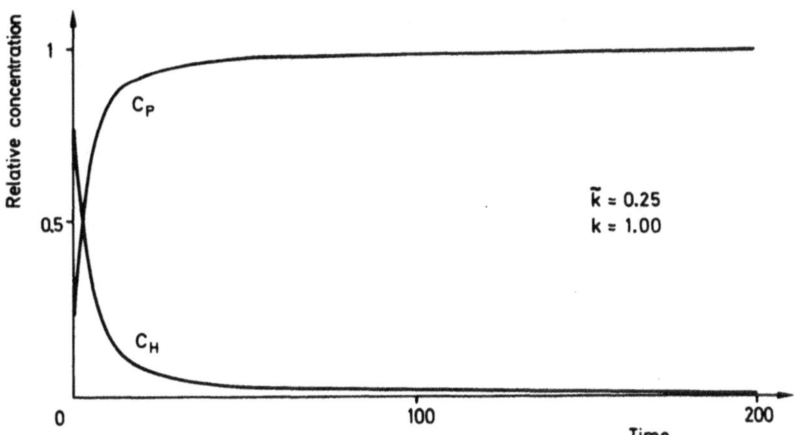
Relative concentration
1
0.5
0
100
200
Time
C_P
C_H
$\tilde{k} = 0.25$
$k = 1.00$

(c)

$$c_A + c_B = c_0 = \text{const.} \tag{9.59}$$

In this case the control parameter φ must satisfy the condition

$$\varphi = \frac{1}{c_0} (\tilde{k}_A c_A^2 + k_A c_A^2 c_B^n + \tilde{k}_B c_B^2 + k_B c_B^2 c_A^m) \tag{9.60}$$

The CP selection equations for a coupled pair of hypercycles are then

$$\dot{c}_A = \tilde{k}_A c_A^2 + k_A c_A^2 c_B^n - \frac{c_A}{c_0} (\tilde{k}_A c_A^2 + k_A c_A^2 c_B^n + \tilde{k}_B c_B^2 + k_B c_B^2 c_A^m) \tag{9.61}$$

and

$$\dot{c}_B = \tilde{k}_B c_B^2 + k_B c_B^2 c_A^m - \frac{c_B}{c_0} (\tilde{k}_A c_A^2 + k_A c_A^2 c_B^n + \tilde{k}_B c_B^2 + k_B c_B^2 c_A^m) \tag{9.62}$$

We can simplify these equations considerably by making $m=n=1$. Catalytic terms of higher order are rare among biological macromolecules, so this simplification appears justified. All terms with $m, n \geq 2$ require either a complex multistep mechanism or a simultaneous encounter between more than two members of the two hypercycles. Both of these are highly improbable. We have already used a similar argument in Section 9.1 in order to limit the treatment of hypercycles to cases of low dimensionality. We can thus also neglect the second-order growth terms for the autocatalysis of the hypercycles H_A and H_B, as long as c_A and c_B are small enough.

With these assumptions we obtain from Eqs. (9.61) and (9.62) the following simplified pair of differential equations:

$$\dot{c}_A = k_A c_A^2 c_B - \frac{c_A}{c_0} (k_A c_A^2 c_B + k_B c_B^2 c_A) \tag{9.63}$$

and

$$\dot{c}_B = k_B c_B^2 c_A - \frac{c_B}{c_0} (k_A c_A^2 c_B + k_B c_B^2 c_A) \tag{9.64}$$

Consider first the long-term behaviour of the system. A fixed-point analysis gives two fixed points $\mathbf{c}^{S_1}$ and $\mathbf{c}^{S_2}$, lying in the apices of the simplex S_2:

$$\mathbf{c}^{S_1} = (c_A, c_B)^{S_1} = (c_0, 0)^{S_1} \tag{9.65}$$

and

$$\mathbf{c}^{S_2} = (c_A, c_B)^{S_2} = (0, c_0)^{S_2} \tag{9.66}$$

The two hypercycles thus cannot co-exist, even though they are catalytically coupled to one another.

Now let us examine the stability of the two fixed points. Since the eigenvalues associated with them are all zero, we must include higher-order terms in our analysis. For $c_A = c_0 - \delta c_A$ we obtain from Eq. (9.62), after a simple rearrangement,

$$\dot{c}_A = \frac{(\delta c_A)^2}{c_0} (c_0 - \delta c_A)^2 (k_A - k_B) \tag{9.67}$$

We see at once that

$$\dot{c}_A > 0 \qquad (\text{for } k_A > k_B) \tag{9.68}$$

and

$$\dot{c}_A < 0 \qquad (\text{for } k_A < k_B) \tag{9.69}$$

The fixed point $\mathbf{c}^{S_1}$ is stable for $k_A > k_B$. Since the Eqs. (9.63) and (9.64) are symmetric with respect to the indices A and B, the fixed point $\mathbf{c}^{S_2}$ is clearly stable when the rate constants fulfil the inequality $k_A < k_B$. The numerical integration of Eqs. (9.63) and (9.64) is shown in Fig. 9.15.

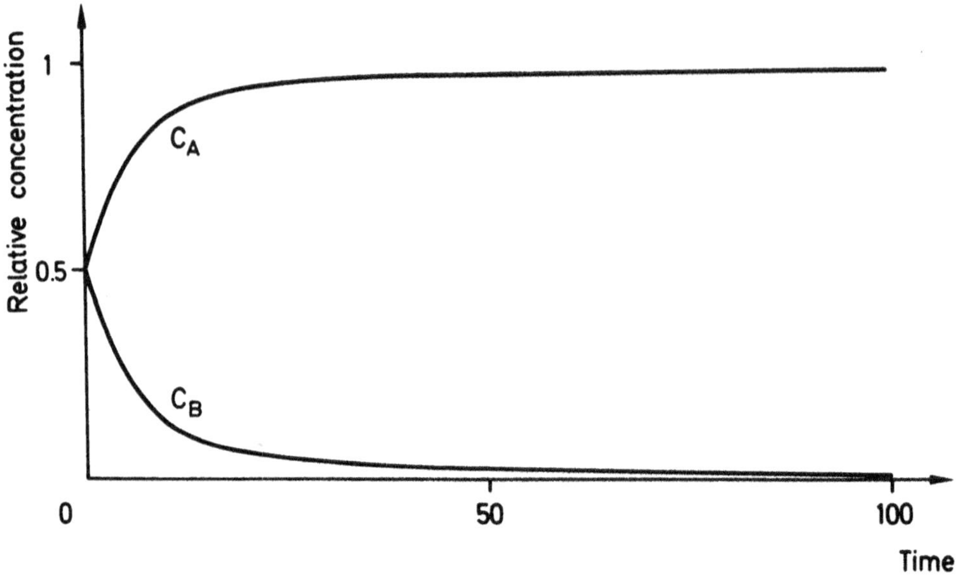

Fig. 9.15. Growth of a catalytically coupled pair of hypercycles under CP conditions. The figure shows the results of numerical integration of the set of Eqs. (9.63) and (9.64). As long as only catalytic terms of the third order appear, co-existence is not possible. Since $k_A > k_B$, H_B dies out and H_A is selected. Although the hypercycles are catalytically coupled, there is selection between them. Both systems start with the same concentration: $c_A(0)=c_B(0)=0.5$; $k_A=2$, $k_B=1$.

We have just considered a system containing third-order coupling terms. A similar investigation of fourth-order coupling has been carried out by M. Eigen and P. Schuster (1979). The analysis shows that in this case the co-existence of two hypercycles again becomes possible. However, the models with catalytic terms of order greater than two are, on account of their small probability of nucleation, of no further interest in connection with the origin of life.

9.4. The Catalytic Hypercycle

Consider the reaction scheme shown diagrammatically in Fig. 9.16. Two classes of biological macromolecules are involved: n polynucleotides I_i and n polypeptides E_j. Each information unit I_i consists of a pair of nucleic acids, which reproduces itself autonomously according to the principle of complementary base-pairing. The reproduction of each $(\pm)$ ensemble I_i is catalysed by the corresponding polypeptide E_{i-1}; the latter is a translation product of the information unit I_{i-1}. The effect of E_n upon I_1 is to close the circle and make the system catalytically coupled. In our terminology, the reaction system in Fig. 9.16 is an elementary hypercycle involving translation and of dimension 2n.

Hypercycles which include translation are sometimes termed *realistic* hypercycles on account of their resemblance to the protein biosynthetic cycle in organisms today.

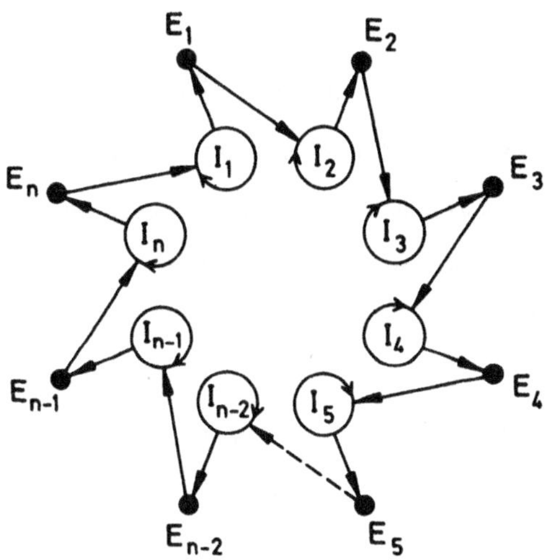

Fig. 9.16. The catalytic hypercycle. The hypercycle with translation is made up of n nucleic acid units I_n and n polypeptides E_n. Each nucleic acid unit consists of a pair of nucleic acid chains of restricted length, which reproduce themselves by cross-catalytic instruction. Let the products of translation of the nucleic acids be catalytically active peptides E which each promote synthesis of the nucleic acid I_{i+1} in the cycle. (After Eigen, 1971.)

We shall see that the catalytic hypercycle really can be regarded as a precursor of the living cell. To do this, we must show in particular that such systems can arise and evolve under prebiotic conditions. However, that question must wait until Section 11.1. First we must examine the general dynamic properties of catalytic hypercycles.

The kinetics of catalytic hypercycles can be described in the following reaction scheme:

(a) Formation of a nucleic acid-protein complex.

$$I_i + E_j \overset{K_i}{\rightleftarrows} I_iE_j \tag{9.70}$$

(b) Protein-catalysed replication of the information unit I_i:

$$I_iE_j + \sum_{\mu=1}^{4} \nu_\mu N_\mu \xrightarrow{k_i} I_i + I_iE_j \qquad \begin{bmatrix} j=i-1+n\delta_{i1} \\ i=1,\dots,n \end{bmatrix} \tag{9.71}$$

(c) Translation of information unit I_i into protein E_i:

$$I_i + \sum_{\mu=1}^{20} \nu_\mu a_\mu \xrightarrow{k_i'} I_i + E_i \tag{9.72}$$

Here ν_μ are the stoichiometric coefficients either of the four classes of ribonucleoside triphosphates N_μ or, as appropriate, of the 20 classes of activated amino acids a_μ. In formulating the kinetic equations we have further assumed that the synthesis of the polypeptides proceeds by means of a general, unspecific translation apparatus equally available to all I_i.

We now introduce a concentration variable z_i for the complex (I_iE_j). In addition we distinguish between the total polynucleotide (x_i) or polypeptide (y_i) concentrations and the concentrations of free (*i.e.*, not present as a complex) polynucleotides (x_i^0) or polypeptides (y_i^0). The law of conservation of matter implies the relations

$$x_i = x_i^0 + z_i \tag{9.73}$$

and

$$y_j = y_j^0 + z_i \tag{9.74}$$

If complex formation between I_i and E_j is very rapid, we can express z_i as a function of the total concentration x_i and y_j. The law of mass-action gives

$$K_i = \frac{z_i}{x_i^0 y_j^0} \tag{9.75}$$

from which follows, using Eqs. (9.73) and (9.74)

$$K_i = \frac{z_i}{(x_i - z_i)(y_j - z_i)} \tag{9.76}$$

This expression leads in turn to a quadratic equation for z_i. Only one of the two solutions is physically meaningful, namely

$$z_i = \frac{x_i + y_j + K_i^{-1}}{2}\left(1 - \sqrt{1 - \frac{4x_i y_j}{(x_i + y_j + K_i^{-1})^2}}\right) \tag{9.77}$$

The assumption of rapidly-reached complexation equilibrium thus allows the reduction of the reaction system in Eqs. (9.70 - 9.73) to two independent variables. Their growth functions are

$$\Gamma_i^{(x)} = k_i z_i \qquad \text{(for the x-variables)} \tag{9.78}$$

and

$$\Gamma_j^{(y)} = k_j' x_j^0 \qquad \text{(for the y-variables)} \tag{9.79}$$

We now place the system under CP constraint and keep, independently of each other, the concentrations of polynucleotide and of polypeptide constant. That is, we make

$$\sum_i x_i = c_0^{(x)} \tag{9.80}$$

and

$$\sum_i y_i = c_0^{(y)} \tag{9.81}$$

The CP selection equations (8.8) thus lead to a system of 2n coupled differential equations:

$$\dot{x}_i = k_i z_i - \frac{x_i}{c_0^{(x)}} \sum_j k_j z_j \tag{9.82}$$

and

$$\dot{y}_i = k_i' x_i^0 - \frac{y_i}{c_0^{(y)}} \sum_j k_j' x_j^0 \tag{9.83}$$

whereby according to Eq. (9.77) z_i is a function of the variables x_i and the association constants K_i.

The system described by Eqs. (9.82) and (9.83) differs from the elementary hypercycles discussed in the above section in that the association constant K occurs as an additional parameter. The influence of this parameter upon the dynamics of the hypercycle has been investigated in detail by M. Eigen and P. Schuster (1979). Mathematical analysis of the differential equations (9.82) and (9.83) shows that K has a critical value at which the dynamic behaviour of the system suddenly changes from one limit cycle to another. This is a manifestation of the so-called Hopf bifurcation. The bifurcation, or branching, appears only in hypercycles of high dimension ($n>4$). For hypercycles with $n \leq 4$, however, the dynamic behaviour is completely equivalent to that of the elementary hypercycles discussed in Section 9.2.

From the numerical investigation of the dynamics of catalytic hypercycles one can draw the following conclusion: all catalytic hypercycles possess an attractor, and this lies in the physically meaningful region of concentration space. This means the stable co-existence of all the components taking part in the hypercycle. The topological structure of the attractor depends upon the concentration variables $c_0^{(x)}$, $c_0^{(y)}$, the association constants K and the dimension n of the system. The attractor is either a stable fixed point or a limit cycle.

The dynamic properties of catalytic hypercycles thus accord qualitatively with those of elementary hypercycles. In particular, a catalytic hypercycle, once selected, will not tolerate (in homogeneous solution) the nucleation and evolution of independent competitors. The genetic code and the chirality of systems which have emerged from a single hypercycle are therefore universal.

10. The Origin of the Genetic Code

The hypercycle is a new principle of self-organization of matter, but does not in itself tell us how it could come into being in the real world of biological macromolecules.

A valuable hint for the solution of this problem is given by the model of the catalytic hypercycle (Section 9.4). However, a prerequisite for the nucleation of a catalytic hypercycle is the existence of an apparatus of translation, implying a pre-defined code. Thus the problem of the origin of the genetic code, along with that of the integration and stabilization of information, is of central importance for an understanding of the origin of life.

The subject of this chapter will therefore be the question of what conditions are necessary for the self-assembly of a prebiotic translation apparatus. We shall start by estimating the probability of spontaneous formation of a code system, and shall then present a model for a primitive mechanism of translation.

10.1. Probability of Nucleation

At present there are basically two hypotheses for the origin of the genetic code:

(1) THE STEREOCHEMICAL HYPOTHESIS. This postulates a specific, chemical interaction which, if it were known, would explain directly the present structure of the genetic code (Woese, 1967 and 1973; Hendry *et al.*, 1981). According to the stereochemical hypothesis each code-word is assigned to a particular amino acid as a consequence of a specific stereochemical interaction between them. However, experiments such as those of C. Saxinger and C. Ponnamperuma (1971) and of M. Raszka and M. Mandel (1972) failed to show the existence of such an interaction. Although the association of

amino acids with certain adaptors (*e.g.*, nucleosides) is not completely unspecific, the specificity observed so far is very weak and does not correspond to the present-day genetic code.

One way or another, if the stereochemical hypothesis were correct, then the assumption that physical conditions made the probability of nucleation of *any* genetic code high enough would suffice to explain the emergence of the present-day genetic code as an inevitable consequence. The stereochemical hypothesis is attractive because of its simple explanation of the universality of the genetic code and on account of the simple mechanism of translation which it suggests: the prebiotic translation of a nucleotide sequence would just be a matter of the amino acids' binding directly or indirectly to their corresponding code-words. This picture reduces the origin of the genetic code to a problem not of logic or organization but of detailed chemical interaction.

(2) THE "FROZEN ACCIDENT" HYPOTHESIS. This is the name given to the current alternative to the stereochemical hypothesis. The "frozen accident" hypothesis starts from much weaker premises by ignoring any specific chemical interaction between amino acids and their related code-words. The genetic code is assumed to have arisen through the establishment of a coincidental pattern of assignment between amino acids and adaptors which has become "frozen in" by evolutionary constraints during the prebiotic nucleation phase. This suggestion thus implies that the present-day code structure is chemically quite arbitrary (Crick, 1968).

Since a stereochemical specificity between amino acids and their adaptors has not yet been demonstrated experimentally, the idea of a "frozen accident" has gained weight in recent years.

The translation apparatus employed by the living cell is astonishingly complex. So let us see how large the probability is for the nucleation of such an apparatus by pure chance.

Consider two sets of μ elements

$$\mathbf{a} = (a_1, a_2, \ldots\ldots, a_\mu) \tag{10.1}$$

and

$$\mathbf{I} = (I_1, I_2, \ldots\ldots, I_\mu) \tag{10.2}$$

We let **a** represent the class of natural amino acids and **I** the class of nucleic acid adaptors or code-words.

We shall calculate the probability that a purely random interaction between the elements of **a** and **I** results in an unambiguous and reproducible assignment. If every element a_i can combine with every element I_j then there are precisely μ^2 possible assignments. Let us assume that each of these correlations has the same probability *a priori*. We now construct a system of notional compartments which are

1	2	3	4	5
a_1I_1 a_1I_1	a_1I_1 a_1I_2	a_1I_1 a_2I_1	a_1I_1 a_2I_2	a_1I_2 a_1I_2
6	**7**	**8**	**9**	**10**
a_1I_2 a_2I_1	a_1I_2 a_2I_2	a_2I_1 a_2I_1	a_2I_1 a_2I_2	a_2I_2 a_2I_2

Table 10.1. Compartmentation of a binary code system. Consider an arbitrary distribution of the assignments a_1I_1, a_1I_2, a_2I_1 and a_2I_2. Let the entire system be divided into compartments which each accommodate on average two translation assignments. There are ten possible ways of occupying a compartment, but only in compartments 4 and 6 do we find unambiguous translation.

just the right size to contain μ assignments. The number of possible compartments is given by the number of different ways of choosing μ elements from a set of μ^2 elements. If we allow repetitions (*i.e.*, an assignment can occur with up to μ-fold redundancy in a single compartment) then the number μ^2 has to be replaced by $\mu^2 + (\mu-1)$, so that there are

$$z_\mu = \binom{\mu^2+\mu-1}{\mu} = \frac{(\mu^2+\mu-1)!}{\mu!(\mu^2-1)!} \tag{10.3}$$

possible different compartments containing μ assignments. The compartments for a binary code system (μ=2) are shown in Table 10.1. If we now examine the contents of each compartment we find that of the z_μ possible occupation patterns only $\mu!$ represent unambiguous code systems. "Unambiguous" in this context means that each amino acid is assigned to one adaptor only, and *vice versa.*

Clearly only a small fraction of all the combinatorially possible correlations can lead to a complete set of unambiguous assignments such as would be needed in a usable genetic code. The probability that one of the $\mu!$ unambiguous sets of assignments (code systems) will be found in a given compartment is

$$\frac{\mu!}{z_\mu} = \frac{(\mu!)^2\ (\mu^2-1)!}{(\mu^2+\mu-1)!} \tag{10.4}$$

However, this is not yet the probability that we are looking for. If a code system which has arisen spontaneously is to be stabilized, then the information carriers must include some which, after translation, *promote the use of the same code system.*

We must therefore find the probability of spontaneous nucleation of a code system which is both unambiguous and which can be used in its own reproduction. To do this we use the same line of argument as in the calculation of the probability given by Eq. (10.4).

The result of the translation of various, different nucleotide sequences is a distribution of polypeptide sequences. Reproducible translation only becomes possible when the translation products work on the information carriers in such a way that the application of the same code system as in the primary round of translation is preferred. We assume that both the nucleic acids and the proteins are present in comparable concentrations. The probability of finding a particular distribution of assignments in a defined element of volume is again given by the reciprocal of z_μ. But we are now looking not for any of the $\mu!$ unambiguous distributions, but for the particular one which was used before for the primary translation.

The probability that a purely random process will lead to an unambiguous *and* reproducible translation system is thus given by the product

$$p = \frac{\mu!}{z_\mu} \cdot \frac{1}{z_\mu} = \frac{(\mu!)^3[(\mu^2-1)!]^2}{[(\mu^2+\mu-1)!]^2} \tag{10.5}$$

Relation (10.5) is particularly important, since it allows a reliable estimation of the upper limit of the probability of nucleation as a function of the dimension μ of the code (see Table 10.2).

μ	2	4	8	20
p	$2\cdot10^{-2}$	$1.6\cdot10^{-6}$	$4\cdot10^{-16}$	$5\cdot10^{-50}$

Tab. 10.2. Probability of the spontaneous formation of an unambiguous and reproducible code system of dimension μ.

Our discussion thus leads to the conclusion that the spontaneous formation of a binary code has a comparatively high probability (see also Section 11.1). The value for nucleation of a four-membered code also lies in a sensible order of magnitude. For an eight-membered code, on the other hand, the value lies well outside the realm of likelihood.

To summarize: It cannot be excluded that the genetic code, as found today in the living cell, arose in a single, random event. The nucleation probabilities, however, suggest strongly that the code originally had a much lower dimension than μ = 20, and only in the course of evolution did it optimize itself and develop to its present-day dimension. The discussion which follows will show that traces of this optimization are to be found in the structure of the present-day genetic code.

A virtue of the "frozen accident" hypothesis is that it dispenses with the assumption of highly specific interactions (direct or indirect) between amino acids and adaptors. In this context we should mention the frequently-heard criticism that the "frozen accident" fails to account for the universality of the genetic code (see *e.g.* Monod, 1971). There must indeed be mechanisms, if the "frozen accident" hypothesis is right, which select, once for all time, an unambiguous and reproducible code system as soon as it appears. However, catalytic hypercycles, discussed at length in the preceeding chapter, have just this property.

10.2. Models of Prebiotic Translation

Protein biosynthesis is today an exceedingly complex process, in which a large number of optimized proteins take part. As we have shown in Chapter 3, it is highly improbable that all these proteins emerged in optimal form from a purely random synthesis during the phase of prebiotic condensation of amino acids. They must therefore have been optimized by mutation and selection, for example in a hypercyclic phase of molecular self-organization.

We must therefore ask whether a primitive translation process could take place without the help of the numerous optimized proteins of the ribosome.

A plausible model for ribosome-free translation of nucleic acids was suggested in 1976 by F.H.C. Crick *et al.* These authors call their idea a speculation, since it proceeds from several unconfirmed assumptions. None the less, we shall consider their hypothesis in detail, since it makes it clear that the catalytic hypercycle is by no means an esoteric or unrealistic concept, but that, on the contrary, it can be built up from a series of plausible single steps at the molecular level. The model for prebiotic translation has therefore at least a certain heuristic value.

The prebiotic backdrop before which Crick *et al.* construct their model for primitive translation consists solely of a distribution of ribonucleic acids and amino acids that are *activated*, *i.e.*, bound to RNA (*cf.* Fig. 2.9). Two demands are now made of the physico-chemical properties of the ribonucleic acids. Firstly, the RNA molecules shall reproduce themselves perpetually by an unspecific copying

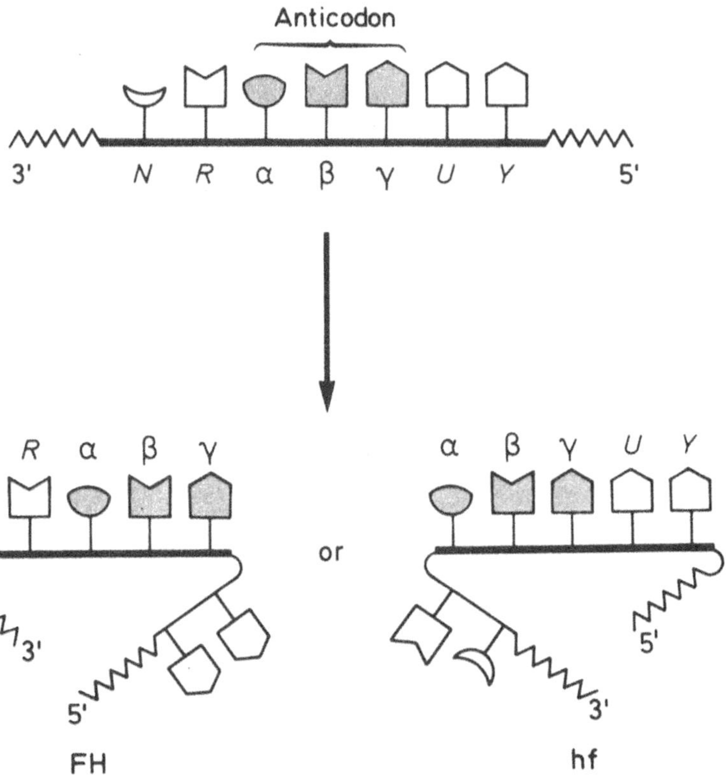

Fig. 10.1. Two stacking configurations of tRNA molecules. It is postulated that the anticodon region of primitive tRNA molecules, containing seven nucleotides, was able to adopt two alternative configurations. (a) In the FH (3′-stacked) configuration, suggested by W. Fuller and A. Hodgson (1967), the five bases at the 3′ end of the anticodon region make up a co-operative unit. (b) In the hf (5′-stacked) configuration, suggested by C. Woese (1970), the five bases at the 5′ end of the anticodon region make up a co-operative unit.

mechanism. Secondly, some of the myriad of RNA molecules shall possess the clover-leaf structure shown by tRNA molecules today (without modified bases). These two demands do not really place any constraint upon the prebiotic scenario, since we already know that the nucleic acids have an inherent capacity for self-reproduction and that all RNA molecules for which high stability and preferential reproduction are important evolve towards tRNA-like structures when put under selection pressure (see also Eigen, 1971).

We shall refer to the primitive tRNA precursors as "adaptors" and proceed from the following three assumptions (Woese, 1970):

(1) The important part of the anticodon region consists of seven nucleotides and possesses two alternative conformations (Fig. 10.1). In one configuration (3′-stacked) the five nucleotides at the 3′ end make up a binding site for mRNA. In the other configuration (5′-stacked) the binding site for mRNA is made up by the five nucleotides at the 5′ end.

(2) An adaptor molecule that is charged with an amino acid adopts the 5′-stacked configuration. An adaptor molecule carrying a growing peptide chain adopts the 3′-stacked configuration. No assumption is made about the configuration of an uncharged adaptor molecule; this could be one or the other or an equilibrium mixture of the two.

(3) Under prebiotic reaction conditions the binding of an adaptor to the mRNA is stable for a finite time.

On the basis of assumptions (1) to (3) a simple mechanism can now be constructed for a nucleic-acid-instructed, ribosome-free protein synthesis, as shown schematically in Fig. 10.2. At some point during the translation (Fig. 10.2a) an adaptor molecule, charged with the nascent peptide chain, binds to the mRNA in the 3′-stacked configuration. The next step in the translation is the binding of a second adaptor molecule, charged with an amino acid and possessing the 5′-stacked configuration (Fig. 10.2b). The growing peptide chain is now transferred to the neighbouring adaptor molecule (Fig. 10.2c). This process induces a configurational change (5′- to 3′-stacked) in the recipient adaptor molecule and thus the displacement of part of the now uncharged adaptor (Fig. 10.2d). The uncharged adaptor falls off the mRNA, so that the initial state (Fig. 10.2a) is restored, except that the site of translation has moved one codon along the messenger. Repetition of steps 10.2a to 10.2d results in the complete translation of the information carrier into the corre-

Fig. 10.2. Suggested model for a prebiotic translation mechanism. (a) The adaptor charged with a growing polypeptide P_n (FH or 3′-stacked configuration) binds to the mRNA with five bases of its anticodon region. (b) The tRNA charged with the next amino acid **a** (hf or 5′-stacked configuration) also binds to the mRNA with five bases of its anticodon region. (c) The growing polypeptide chain is transferred to the amino acid **a** on the newly-arrived adaptor, with the formation of a peptide bond between P_n and **a** , leading to the polypeptide P_{n+1}. (d) The adaptor charged with P_{n+1} changes its configuration from hf to FH. The uncharged adaptor is now bound by only three of its bases to the mRNA. The binding is therefore destabilized and the tRNA dissociates from the mRNA. After step (d) the initial situation is restored.
Successive iteration through steps (a) to (d) results in the synthesis of the protein for which the mRNA codes. Frame-shifts during translation are impossible, so that the translation takes place unambiguously and reproducibly. (After Crick *et al.*, 1976.) ▶

sponding protein. Although the reading frame in this translation model is five bases long, the translation takes place according to a triplet code, the difference being accounted for by the double use of the codon bases.

Suppose that the above mechanism was indeed used during the prebiotic phase of instructed protein synthesis. What consequences would this have concerning the structure of the primitive code?

To answer this question we take a closer look at the structure of tRNA. As an analysis by B.G. Barrell and B.F.C. Clark (1974) has already shown, the anticodon region of all known tRNA sequences has the pattern (see also Fig. 10.3)

$$3' \ldots\ldots\ldots\ldots\ldots \mathrm{NR}\,\alpha\beta\gamma\,\mathrm{UY} \ldots\ldots\ldots\ldots\ldots 5'$$

where the symbols have the meanings:

- $\alpha\beta\gamma$ = anticodon sequence with γ as the "wobble" position (Section 2.3)
- N = any of the four ribonucleotides bases (A, U, G, C)
- R = purine base (A or G)
- Y = pyrimidine base (U or C)

We now make the plausible assumption that the uniform pattern in the anticodon region of present-day tRNA's was also a feature of the tRNA precursors in the prebiotic phase. In addition we shall assume:

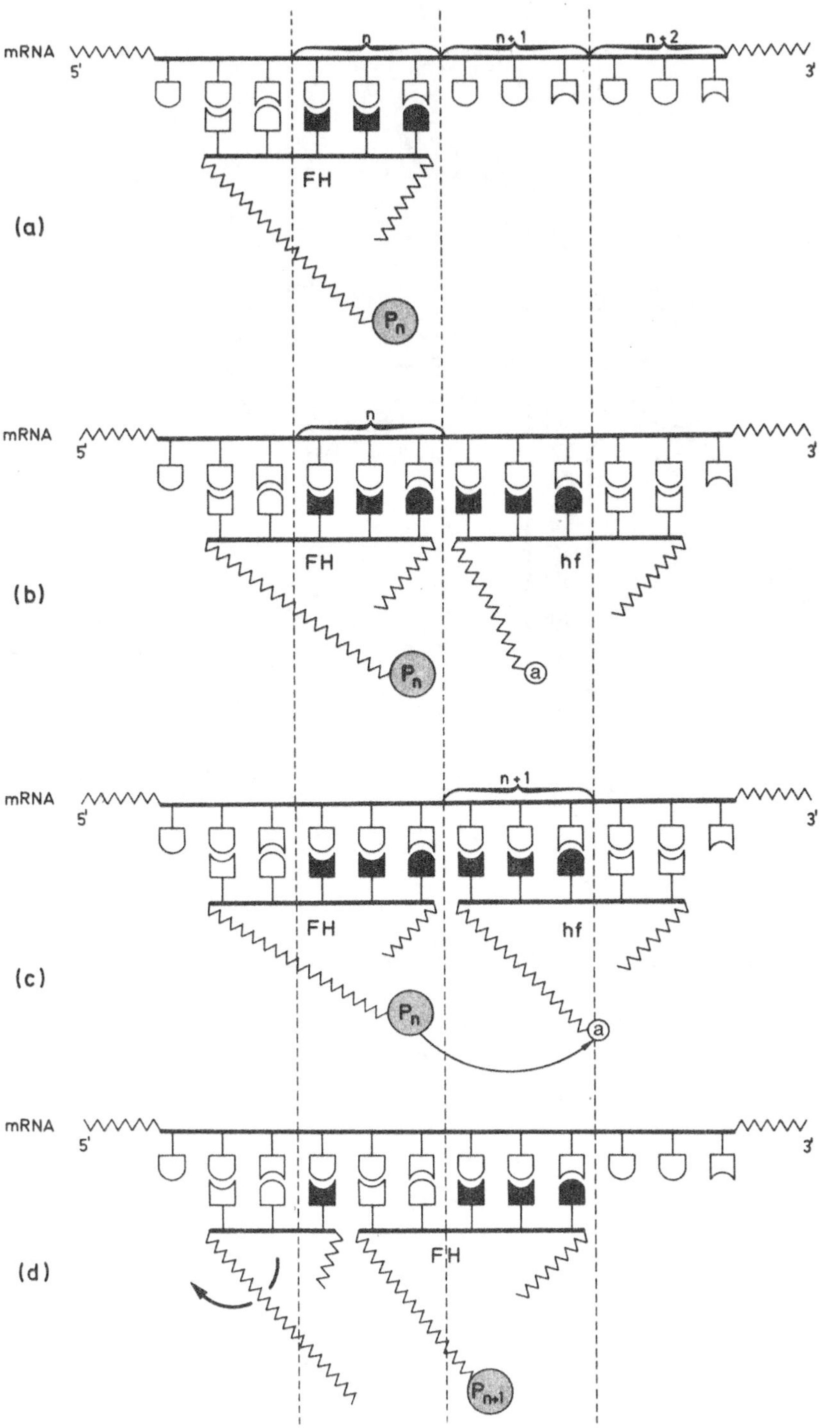
mRNA
5'
3'
n
n+1
n+2
FH
(a)
Pn
mRNA
5'
3'
n
FH
hf
(b)
Pn
a
mRNA
5'
3'
n+1
FH
hf
(c)
Pn
a
mRNA
5'
3'
FH
(d)
Pn+1

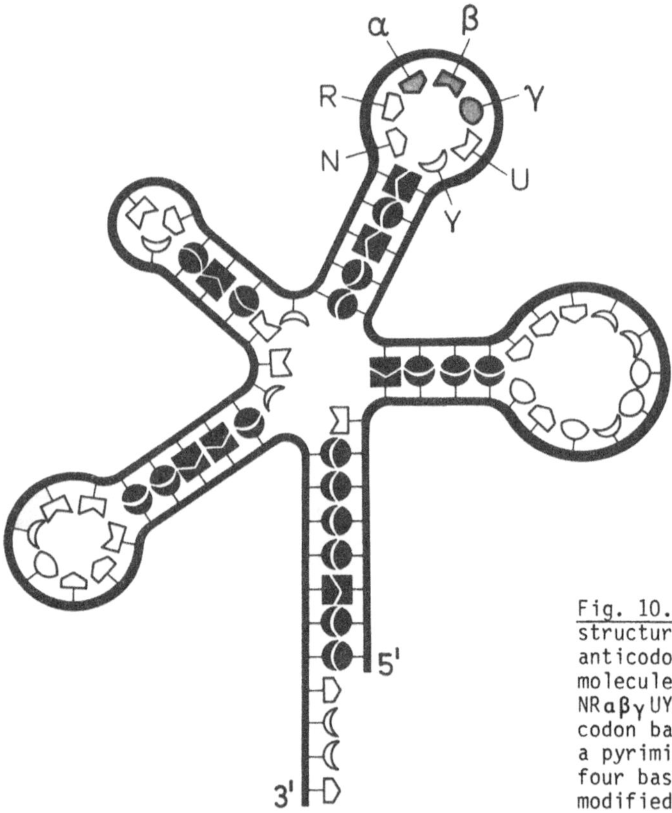

Fig. 10.3. Representative clover-leaf structure of a tRNA molecule. The anticodon regions of present-day tRNA molecules have a uniform pattern NRαβγUY, where α, β, γ are the anticodon bases, R is a purine base, Y is a pyrimidine base and N is any of the four bases (sometimes in a chemically modified form).

(a) In prebiotic adaptors the pyrimidine base Y near the anticodon was always a U (never a C).

(b) All bases in the anticodon region were potential "wobble" bases, *i.e.*, GU pairs were allowed.

Assumption (a) restricts still further the variety of sequences in the anticodon region, leading to

$$3' \ldots\ldots\ldots\ldots\ldots NR\,\alpha\beta\gamma\,UU \ldots\ldots\ldots\ldots\ldots 5'$$

the consistent assumption of which also narrows down the range of possible prebiotic

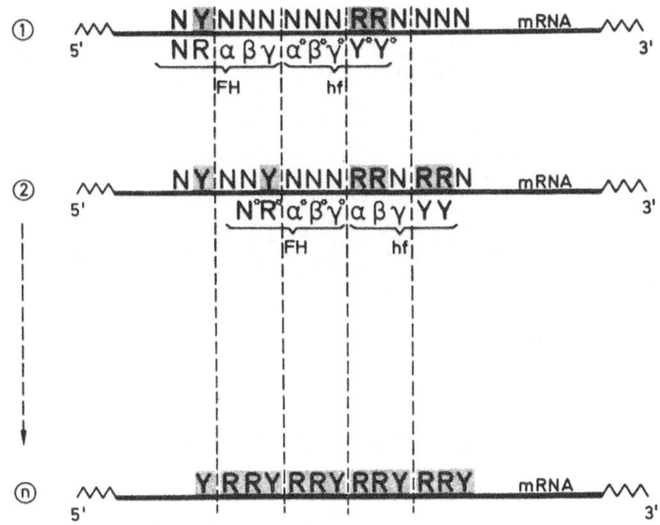

Fig. 10.4. Derivation of the RRY pattern for prebiotic mRNA molecules. Iterative application of the translation mechanism shown in Fig. 10.2 implies an RRY pattern for the prebiotic code.

code structures. The adaptors can only bind in both 3′- and 5′-stacked configurations if the sequence in the anticodon region has a certain degree of symmetry. This means in turn that the primordial mRNA must have possessed a certain kind of repeating pattern. As Fig. 10.4 shows, the repeated alternation between the 5′- and 3′-stacked configurations required by the translation mechanism enables us to deduce the code pattern

5′.....RRY RRY RRY RRY RRY RRY RRY.....3′

If the assumption that all seven bases in the anticodon region bind during a translation step is weakened to embrace only five bases, then, as Fig. 10.5 shows, a more variable code pattern follows:

5′.....RNY RNY RNY RNY RNY RNY RNY.....3′

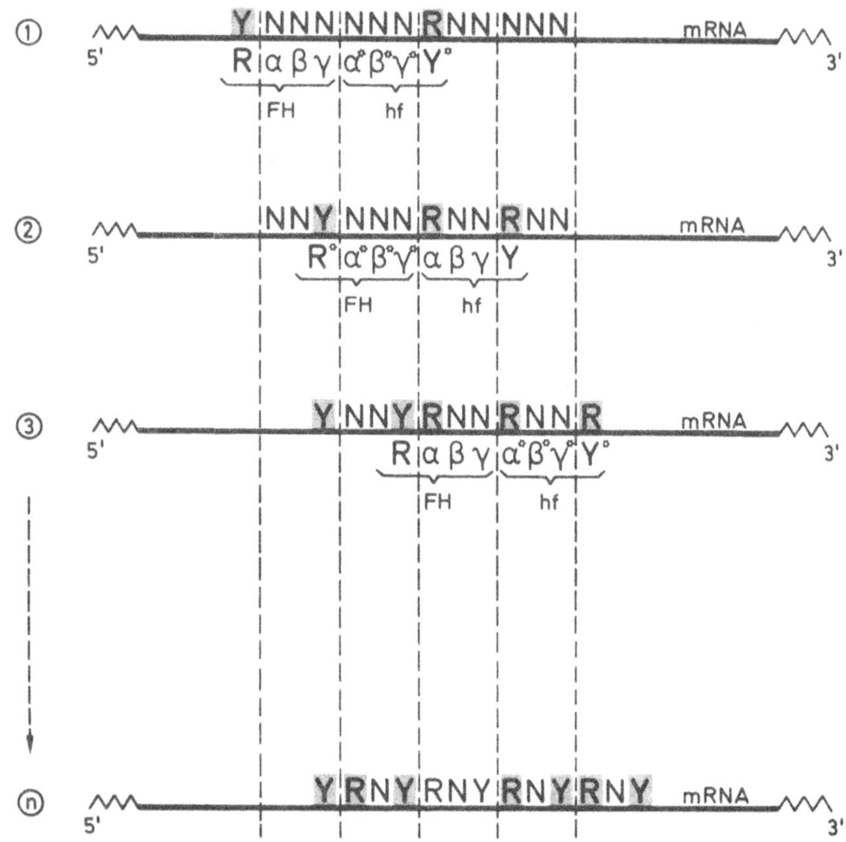

Fig. 10.5. Derivation of the RNY pattern for prebiotic mRNA molecules. The assumption that not five but only four of the bases in the anticodon region bind mRNA leads by way of iterative application of the translation mechanism of Fig. 10.2 to the deduction of an RNY pattern for the prebiotic code.

which reduces to the previous one in the special case of N=R. There are several logical and physical reasons to suppose that an RNY code is more advantageous in evolution than a RRY code. Firstly, the cross-catalytic reproduction of nucleic acids leads automatically in the case of an RNY code to the same code pattern in the complementary strand (see also Section 11.1). Secondly, the RNY pattern is less restrictive than the RRY pattern. It allows more variety in the mRNA sequences and thus possesses more potential for evolution. Thirdly, it does not require the assumption that "wobble" pairing was allowed at each codon position. Further advantages of the RNY code pattern are discussed by M. Eigen and P. Schuster (1979).

At this point we should emphasize the *general* merits of the translation model which we have described:

(i) Protein synthesis is a nucleic-acid-instructed polymerization reaction which does not require a complex apparatus of translation.

(ii) Translation takes place over the entire mRNA chain length.

(iii) The mechanism of translation excludes errors due to phase shift in the decoding frame.

The advantages of unambiguous and reproducible translation are so great that they must have had an enormous stabilizing influence upon the evolution of the early code structure. We should therefore look to see whether present-day information carriers bear traces of the primitive code.

Sequence analysis to this end was first carried out by J.C.W. Shepherd (1981a,b). The aim of his work was to find correlations between the positions of purines (R) and pyrimidines (Y), especially those which might reveal traces of the RNY pattern.

We shall demonstrate the principle of these investigations using an artificial nucleotide sequence as an example. Take the sequence

5′.....AACGUCGGCAUUACUAAUGUUGACGGC.....3′

Since we are interested in correlations between R and Y, we first transcribe the sequence into the appropriate symbolism

5′.....RRYRYYRRYRYYRYYRRYRYYRRYRRY.....3′

and then look for any correlation between the relative positions of R and Y. For example, let us look for correlations of the type $YR(N)_nYY$. We first find all the YR pairs and mark them, *i.e.*,

5′.....RRYRYYRRYRYYRYYRRYRYYRRYRRY.....3′

We then count the frequency with which a YY combination occurs as a function of the number n of bases separating it from the marked YR combinations. The frequency we denote with z. In accordance with the present-day direction of translation we only look for correlations in the 5′ to 3′ direction. As can easily be checked by counting, our model sequence contains exactly four incidences of the combination YRYY. There are no incidences of YR(N)YY or of $YR(N)_2YY$. The combination $YR(N)_3YY$,

GGC gly	GAC asp	GUC val	GCC ala
GGU gly	GAU asp	GUU val	GCU ala
AGC ser	AAC asn	AUC ile	ACC thr
AGU ser	AAU asn	AUU ile	ACU thr

Table 10.3. The primordial code as derived from an RNY code pattern. (According to Eigen and Schuster, 1979.)

however, occurs three times in all. Thus the combination $YR(N)_nYY$ occurs only if n = 0, 3, 6,..., and this is of course because our model sequence conforms perfectly to the pattern RNY.

If our artificial sequence were to undergo an arbitrary number of mutations, the sharp correlation would become blurred after a while, with an increasing frequency of incidence for n = 1, 2, 4, 5,... at the expense of n = 0, 3, 6,... . We should therefore expect the prebiotic code pattern to have disappeared completely on account of innumerable mutations in the course of evolution. It is all the more surprising

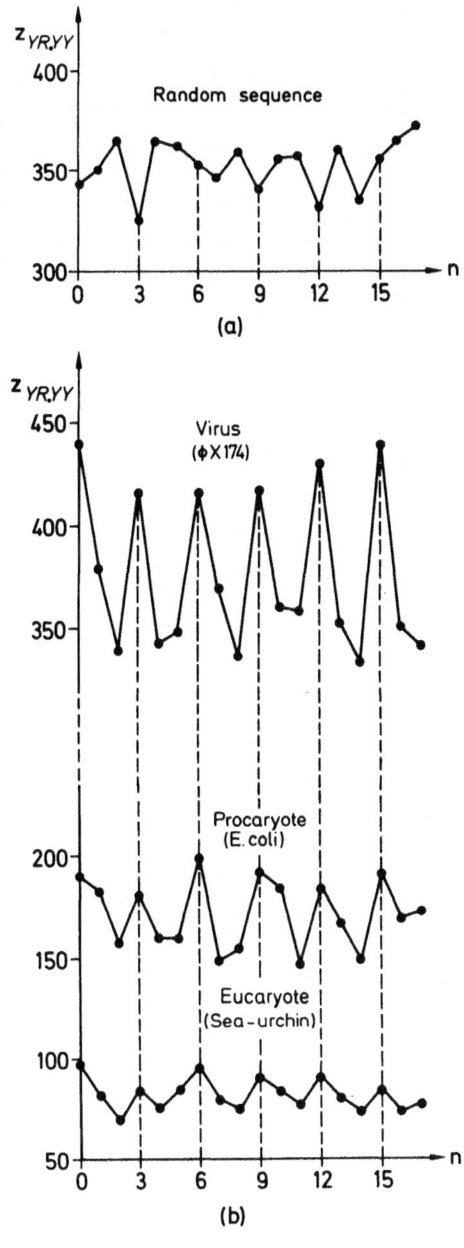

Fig. 10.6. Rhythmic correlations in the purine-pyrimidine distribution in present-day mRNA molecules. The total number $z_{YR.YY}$ of all the purine-pyrimidine combinations of the type YR followed n bases later by the combination YY was determined. The graphs thus show the frequences of occurrence of combinations $YR(N)_nYY$ (abbreviated: YR.YY) as a function of n (a) for a random sequence and (b) for various natural RNA sequences: mRNA from virus ϕX174; ribosomal gene cluster from *E.coli*; sea-urchin histone genes. Similar correlations are found *e.g.* for YR.YR. (After Shepherd, 1981a,b.)

Codon		Amino acid	Abundance of natural amino acids	
			Simulated prebiotic synthesis Yield [μM]	Murchison meteorite μg/g meteorite
1	GG_U^C	gly	440	6
	GC_U^C	ala	790	3
2	GA_U^C	asp	34	2
	GU_U^C	val	19.5	2
3	AG_U^C	ser	5.0	-
	AC_U^C	thr	1.6	-
4	AU_U^C	ileu	4.8	-
	AA_U^C	asn	-	-

Table 10.4. Comparison between primordial codon and amino-acid frequencies. The codon frequency in primitive translation (GNC > ANU > others) was deduced from known base-pair stabilities of the nucleic acids (for details see Eigen and Schuster, 1979).

The amino-acid distribution is the result of experiments to simulate chemical evolution (Miller and Orgel, 1974) and of the chemical analysis of the Murchison meteorite (Kvenvolden *et al.*, 1970 ; Oró *et al.*, 1971; Cromin and Moore, 1971).

that, even today, regular purine-pyrimidine correlations should be clearly visible in natural nucleotide sequences (Fig. 10.6). The correlations are indeed very marked, and support the assertion that the prebiotic code was based on an RNY pattern.

It is worthy of note that this periodicity appears with particular emphasis in sequences which code for proteins. In fact the R-Y correlation allows the determination of the frame of translation without previous knowledge of the positions of the stop and start signals. This is true of genes for proteins from widely-differing organisms (viruses, procaryotes and eucaryotes).

The code pattern RNY deduced from the prebiotic translation model automatically lays down a structure for the prebiotic code. It has 16 code-words, shown in Table 10.3.

Using arguments based on relative base-pair stabilities M. Eigen and P. Schuster (1979) have deduced an order of the frequencies of appearance of the 16 code-words under prebiotic conditions (Table 10.4). Since the GC base pair is stabler than the AU pair one can well imagine that codons of the form GNC should have played a more important part in primitive translation (*i.e.*, in the interaction between codons and anticodons) than those of the form ANU. For comparison, Table 10.4 also shows the frequency distribution of natural amino acids found in simulation experiments under supposedly prebiotic reaction conditions (see Miller and Orgel, 1974). It is conspicuous that the first four codons in the Table are precisely those attributable to the amino acids presumed to have been most common during the phase of chemical evolution. This seems to support the "frozen accident" hypothesis of the origin of the code. The other four codons are also assigned to amino acids which (with the exception of asparagine) were also detected in significant yield in the simulation experiments. The amazing correspondence between the codon frequency deduced from theory and the amino-acid frequency found in experiments provides, looking back, a further reason to incorporate the translation model of Crick *et al.* at least tentatively into our picture of the origin of life.

11. The Evolution of Hypercycles

The information-integrating properties of the hypercycle make it almost inevitable that something of this kind was the forerunner of the protocell. However, as we have seen in the previous chapter, there are considerable problems connected with the question of how hypercycles themselves arose.

On the one hand, an optimized translation system requires so many specific coupling factors that only a hypercycle of high dimension could support such a system. On the other hand, the probability of nucleation for an optimized translation apparatus is, on account of the large number of factors involved, vanishingly small (see Section 10.1).

We are therefore left with the question of whether there is a mechanism of optimization by which a primitive hypercycle, of low dimension, can evolve gradually into a catalytic hypercycle of high dimension.

To begin with we shall formulate an evolutionary principle for hypercycles in general terms. We shall then deal briefly with the questions of whether and to what extent the traces of such an optimization may be found in the components of the present-day translation apparatus. In the second part of this chapter we shall trace the path from the catalytic hypercycle *via* compartmentation and individuation to the proto-cell. Since our understanding of compartmented, catalytic hypercycles is at present very much in a state of flux, we shall have to restrict our discussion to the demonstration of a few basic principles.

11.1. Self-Organization of Catalytic Hypercycles

The advantage of the hypercycle is that the individual information units of the system promote each other's growth, which gives the entire system an enormous advantage when it is put under selection pressure. In the catalytic hypercycle the promotion of growth takes place by means of proteins, which are the translation products of the nucleic acids in the hypercycle.

We shall first estimate the probability with which a hypercycle can arise spontaneously. For this purpose we consider the formation of an information carrier I_i, described by the rate equation

$$\dot{x}_i = k_i x_i x_{i-1} \tag{11.1}$$

and estimate the number of catalytically active molecules which would be needed if the information carriers were to reduplicate themselves with a half-time of, for example, one day. We replace the differential increments in Eq. (11.1) by differences, *i.e.*,

$$\dot{x}_i = \frac{dx_i}{dt} = k_i x_i x_{i-1} \approx \frac{\Delta x_i}{\Delta t} \tag{11.2}$$

and insert values of $\Delta x_i = 2x_i$ and $\Delta t = 9 \cdot 10^4$s. For k_i we shall assume the highest conceivable rate of reproduction, that is, the rate of a diffusion-controlled reaction. For macromolecules this lies in the order of magnitude

$$k_\mathcal{D} = 10^8 \text{ l/mol} \cdot \text{s} \tag{11.3}$$

Using Eq. (11.2) we obtain for x_{i-1} the relation

$$x_{i-1} = \frac{2x_i}{k_\mathcal{D} x_i \Delta t} = \frac{1}{4.5}\, 10^{-12} \text{ mol/l} \tag{11.4}$$

Fig. 11.1. Quasi-continuous evolution of a several-membered hypercycle with translation. The scheme shows the development of hypercyclical self-organization, starting with a single quasi-species consisting of a distribution of (G,C)-rich RNA sequences. (a) Commencement of translation in the form of a statistical assignment of amino acids and anticodons. As long as there is no positive effect of the translation product upon the information carrier, the translation is irreproducible and is subject to uncontrolled fluctuations. (b) An RNA whose reproduction is stimulated by its translation product becomes the dominant sequence within the quasi-species. (c) Mutant I_2, closely related to I_1, leads to a similar translation product. (d) Two closely-related mutants I_1 and I_2, which code for their own replicases E_1 and E_2, can promote each other's growth on account of their similarity. (e) The integration principle of (c) and (d) can be generalized. (f) Every mutant I_3 which offers an advantage can be integrated into the hypercycle.
Certain viruses are known to reproduce themselves according to the coupling scheme (b); details of this are given in Part IV. (After Eigen and Schuster, 1979.) ▶

Since there are $6.2 \cdot 10^{23}$ (Avogadro's number) solute molecules in one litre of a 1-molar solution, x_{i-1} has a value around 10^{11} molecules/litre. Therefore a litre of a prebiotic reaction mixture would have had to contain at least 10^{11} identical copies of a catalytic reaction partner I_{i-1} in order for the number of information units I_i to double in a day. Consequently, the chance nucleation of a functional, several-membered hypercycle would require correlated fluctuations of unusual size.

A fluctuation sufficiently large to lead to the formation of a catalytic hypercycle is thus extremely improbable. However, as argued in Chapter 3, we wish to avoid postulating singular, random events in explaining the origin of life. It is therefore necessary to look for a systematic path which could have led to the self-organization of prebiotic information carriers into hypercycles. For this purpose we shall again make use of the concept of a quasi-species.

Assume the existence of two mutants (I_1 and I_2) of an RNA quasi-species, both present in reasonably large number, whose (+) and (-) strands serve as adaptors for two pairs of amino acids (a_{+1}, a_{-1}, and a_{+2}, a_{-2}) and which can be translated into proteins made up only of these four amino acids (Fig. 11.1a). If a mutation confers upon the translation product of one of the $I_\pm$ ensemble a catalytic function (such as replicase activity) which promotes the reproduction of the corresponding information unit, the result is the coupling scheme shown in Fig. 11.1b.

Since I_1 and I_2 are closely-related mutants we can assume that they are very similar, and that they and their translation products stimulate each other

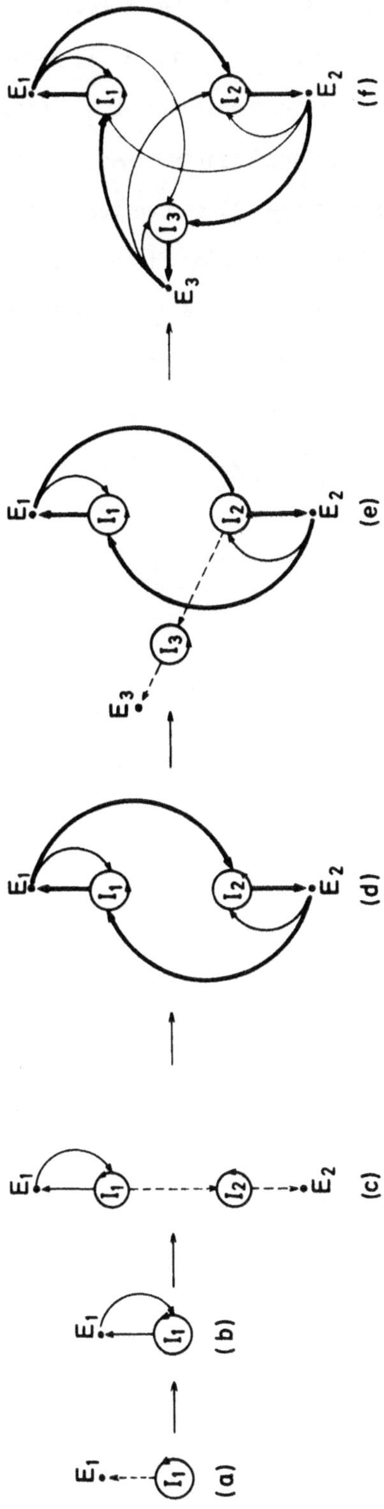

E1
I1
E2
I2
E3
I3
(a)
(b)
(c)
(d)
(e)
(f)

cross-catalytically. The result is the reaction pattern shown in Fig. 11.1d. Whether the cross-catalytic coupling of Fig. 11.1d becomes established is determined by relations between the rate constants k_{ij} (Figs. 11.2 and 11.3): if the cross-catalytic rate constants are greater, then the new coupling pattern is stable. Further development along these lines can turn the original quasi-species step by step into a hypercycle of high dimension (Fig. 11.1e and 11.1f). We shall now demonstrate this using as an example a basic, two-membered hypercycle.

Under CP constraint the growth equations for the information carriers are

$$\dot{x}_i = \sum_{j=1}^{2} k_{ij}x_ix_j - \frac{x_i}{c}\sum_{k=1}^{2}\sum_{m=1}^{2} k_{km}x_jx_m \qquad (i = 1,2) \tag{11.5}$$

with

$$c = x_1+x_2 \tag{11.6}$$

The set of differential equations (11.5) has three fixed points:

$$\mathbf{x}^{s_1} = (c,0) \tag{11.7}$$

$$\mathbf{x}^{s_2} = (0,c) \tag{11.8}$$

$$\mathbf{x}^{s_3} = (k_{22}-k_{12},\ k_{11}-k_{21})\,\frac{c}{k_{11}-k_{21}+k_{22}-k_{12}} \tag{11.9}$$

with the corresponding eigenvalues

$$\omega_1 = (k_{21}-k_{11})c \tag{11.10}$$

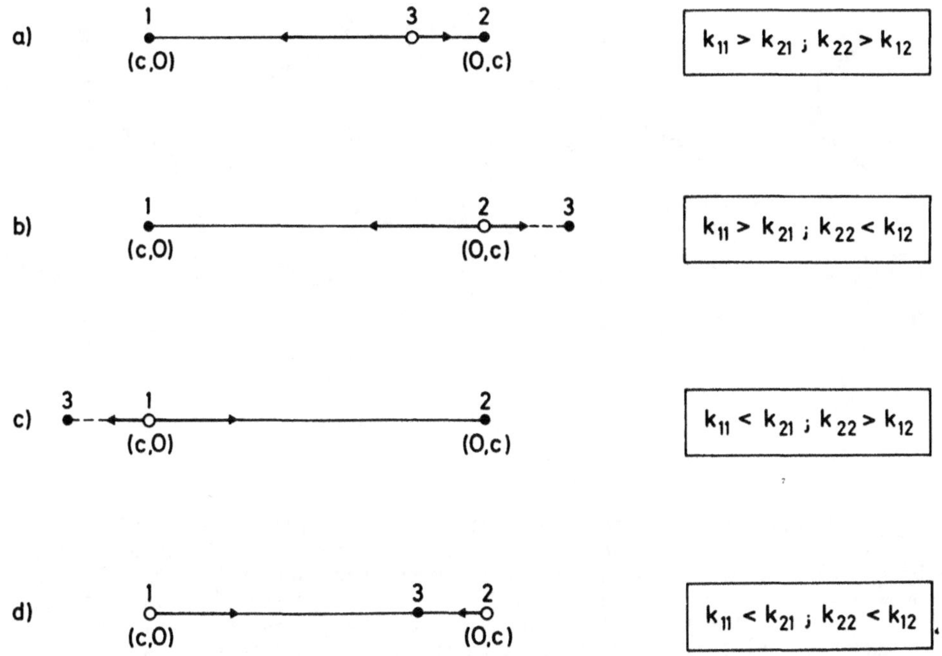

Fig. 11.2. Fixed-point diagram of Eqs. (11.5) for different relationships between the rate constants k_{ij}. (After Eigen and Schuster, 1979.)

$$\omega_2 = (k_{12}-k_{22})c \tag{11.11}$$

$$\omega_3 = \frac{(k_{11}-k_{21})(k_{22}-k_{12})}{k_{11}-k_{21}+k_{22}-k_{12}} \tag{11.12}$$

Four cases may be distinguished, according to the relative magnitudes of the rate constants (Fig. 11.2), and extension of the treatment to include the two-dimensional catalytic hypercycle, as illustrated in Fig. 11.3, leads to the same conclusions:

(a) $k_{11} > k_{21}$ and $k_{22} > k_{12}$: Since I_1 and I_2 are both promoted hypercyclically by their own translation products, the result is sharp selection between I_1 and I_2 (Fig. 11.2a).

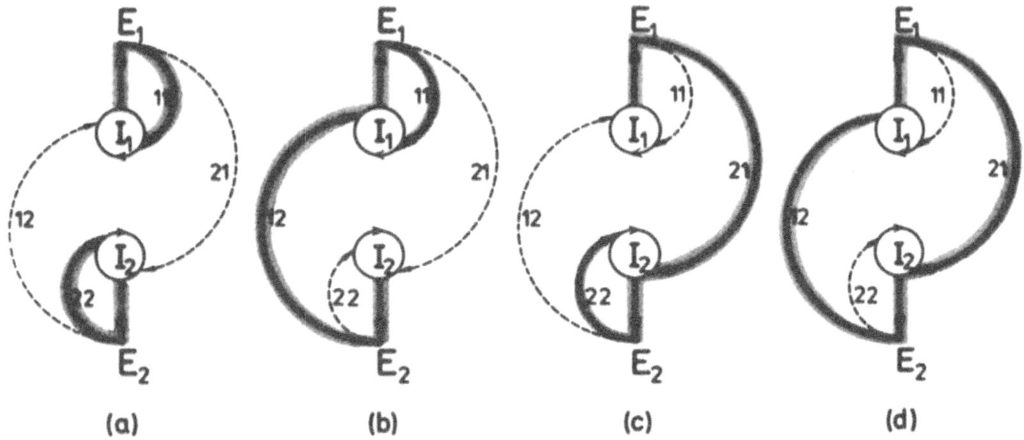

Fig. 11.3. Evolution of a two-membered, catalytic hypercycle. The coupling along the thick lines is stronger than along the broken lines. Four kinds of coupling are possible: (a) E_1 promotes the growth of I_1 (more than that of I_2) and E_2 promotes the growth of I_2 (more than that of I_1). (b) The growth of I_1 is promoted (more strongly than that of I_2) by both E_1 and E_2. (c) The growth of I_2 is favoured by both E_1 and E_2. (d) The growth of I_1 is favoured by E_2 and that of I_2 by E_1.
Only coupling pattern (d) is stable in the long term, allowing the concomitant growth of I_1 and I_2. This behaviour is seen even when the differences between coupling constants are extremely small. (From Eigen and Schuster, 1979.)

(b) $k_{11} > k_{21}$ and $k_{22} < k_{12}$: The growth of I_1 is favoured over that of I_2, so I_1 is selected (Fig. 11.2b).

(c) $k_{11} < k_{21}$ and $k_{22} > k_{12}$: The growth of I_2 is favoured over that of I_1, so I_2 is selected (Fig. 11.2c).

(d) $k_{11} < k_{21}$ and $k_{22} < k_{12}$: Mutual, hypercyclical stabilization of I_1 and I_2 takes place (Fig. 11.2d).

The dynamic behaviour of the hypercycle shown in Fig. 11.3 is manifested even if the differences between the rate parameters are very small. The system therefore reacts very sensitively to any mutation which lifts the degeneracy of the coupling constants k_{ij}.

It has been shown by M. Eigen and P. Schuster (1979) that this principle can be extended by the method of induction to cover the n-dimensional case for all $n > 2$.

We may therefore conclude that hypercyclical organization of any degree can emerge from a quasi-species by mutation and selection.

This evolutionary principle is based on the assumption that the primitive gene had a double function: that of adaptor and that of messenger. We can therefore ask whether the sequences of present-day nucleic acids contain any traces of the early nucleation phase of the translation apparatus, and, if so, whether the sequence of bases in a primitive gene can be reconstructed. We are encouraged to believe that this might be so by the way in which the protein cytochrome c allows us to reconstruct the history of its evolution by comparative sequence analysis (Fig. 3.2). Sequence analysis has also revealed traces of the primordial code pattern in present-day mRNA (Section 10.2).

It is a far harder task to reconstruct the nucleotide sequence of the primitive gene. First of all, we must extrapolate our knowledge much further back than phylogenetic trees enable us to do. Secondly, we do not know which class of present-day nucleic acids is the most direct descendent of the primitive gene. There is here a gap in our empirical knowledge which can be closed only by theory.

The main properties which theory leads us to postulate for the primordial gene are summarized in Table 11.1.

(1) As we have seen in Section 10.2, the formation of a translation apparatus must have been preceded by a structural evolution of nucleic acids. In this early phase of evolution, selection pressure was exerted directly on the information carriers and not indirectly, by way of the gene product. The selective evaluation process must therefore have been based on the phenotypic properties of the nucleic acids themselves. RNA molecules are particularly suited for this kind of selection, since they can - in contrast to DNA - adopt complex tertiary structures and thus show differentiated phenotypic properties (*cf.* Fig. 13.4). So the primitive gene was probably encoded in an RNA molecule.

(2) All RNA molecules for which preferential reproduction and at the same time high stability are important evolve towards symmetric and stable folded structures. This is a consequence of the law of optimization (Eq. 5.107), according to which the

Polynucleotide	Primitive gene (proposed)	Present-day tRNA (observed)
Class and chain length (1,5)	RNA 50 - 100 nucleotides	RNA typically 76 nucleotides, some modified
Composition (4)	G,C >> A,U	G,C : A,U = 2 : 1
Pattern of primary structure (3)	$(RNY)_n$	Relative codon frequencies: 10 RNY 6 RNR 5 YNR 3 YNR
Secondary structure (2)	Symmetry and stable folding	Symmetry and stable folding

Table 11.1. Postulated properties of the primitive gene and observed properties of present-day tRNA molecules. (1)-(5): see accompanying text. (After Eigen and Schuster, 1979.)

overall growth parameter of a (±) ensemble is determined by the geometric mean of the growth parameters of the individual components. Optimization of growth properties under selection pressure will therefore inevitably lead to symmetric or to near-symmetric structures, since in a symmetric structure an advantageous mutation automatically affects the phenotypic properties of (+) and (-) strands to an approximately equal extent. This prediction has been confirmed by experiment (see Section 13.2): RNA molecules optimized in "evolution experiments" show typically a high degree of internal symmetry (Fig. 13.7).

(3) The translation mechanism discussed in Section 10.2 suggests that a repeating RRY or RNY sequence would make a messenger suitable for unambiguous translation under prebiotic conditions. Of these two possible code patterns, the RNY code has the advantage that it is symmetric in (+) and (-) strands. The pattern of the (+) strand

$$5'..........RN_+Y\ RN_+Y\ RN_+Y\ RN_+Y\ ....3'$$

is retained in its copying product, the (-) strand:

$$3'..........YN_-R\ YN_-R\ YN_-R\ YN_-R..........5'$$

Thus in the 5′ to 3′ direction both strands possess the same code pattern RNY. This means (i) that both (+) and (-) strands can act as messengers and (ii) that internal molecular symmetry leads to comparable secondary and tertiary structures and thus to similar phenotypic properties and selection values (*cf.* (2) above) for both strands.

The RNY code thus has a decisive advantage for the self-organization of catalytic hypercycles: all of the RNA components needed for the evolution of a translation apparatus can, according to the evolutionary principle described above, develop from a single RNA quasi-species and undergo mutual stabilization.

(4) Measurements of the stability and lifetime of GC and AU pairs imply that in the early phase of primitive translation (G,C)-rich sequences had the advantage over

(A,U)-rich sequences (see Section 10.2).

(5) Under conditions of enzyme-free replication the information storage capacity of RNA molecules was limited to a length of 50-100 nucleotides (Table 7.3).

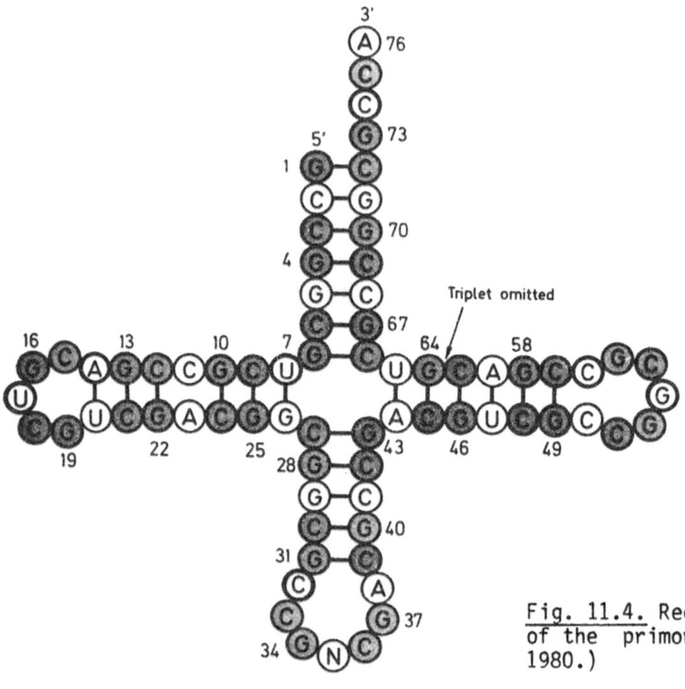

Fig. 11.4. Reconstruction of the sequence of the primordial gene. (From Eigen, 1980.)

This upper limit is therefore something like the length to be expected of the primitive gene, before it was incorporated into a catalytic hypercycle.

As Table 11.1 shows, there is rather close agreement between the structural features postulated for the primitive gene and those observed in present-day tRNA molecules, an agreement which fits in very well with the evolutionary model for catalytic hypercycles discussed at the beginning. Thus it is likely that the primitive gene was a tRNA-like molecule, and it is therefore reasonable to look for traces of its sequence in present-day tRNA's.

M. Eigen and R. Winkler-Oswatitsch (1981a,b) have attempted a reconstruction of the nucleotide sequence of the primitive adaptor - and thus, perhaps, of the

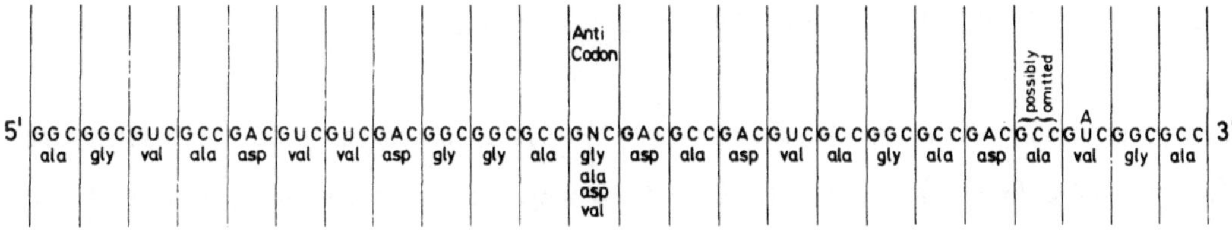

Fig. 11.5. Translation product of the reconstruction of the primitive gene. The translation of the primitive sequence in Fig. 12.4 leads to a primitive protein containing 24 amino-acid residues. Partial blurring of the master sequence and uncertainty of the original nature of the present-day "constant" positions leave the reconstructed primitive protein with several variable positions. (After Eigen, 1980.)

primitive gene - by comparative sequence analysis of present-day tRNA's (Fig. 11.4). The translation of this sequence into a primitive protein is shown in Fig. 11.5.

Since the criteria for the classification of the primitive gene have until now received only limited verification, attempts at its reconstruction must be accorded the status of hypotheses. Nevertheless, the reconstruction of a primitive gene or protein provides a milestone in the development of our picture of prebiotic evolution, since it can be tested experimentally: it is today a perfectly feasible task to synthesize proteins of defined sequence and to examine directly their functional properties.

Recently M.W. Kilpatrick and R.T. Walker (1981) have sequenced the methionine tRNA from the bacterium *Thermoplasma acidophilum*, a member of the Archaebacteria (considered to be the oldest family of organisms, see Fig. 1.1). They found no close sequence homology between this tRNA and any other methionine tRNA so far sequenced but a surprisingly high degree of homology ($\approx 90\%$) with the nucleotide sequence proposed for the ancestral master sequence depicted in Fig. 11.4.

11.2. Compartmentation and Individuation of Hypercycles

Every organism with an autonomous metabolic system possesses a cellular structure. The *compartmentation* and *individuation* of hypercyclically organized systems therefore seems to be a necessary stage in their evolution. Even if this question cannot be treated in detail using the theory we have set up so far, it does not seem to present any fundamental difficulties or contradictions.

There are several plausible reasons why compartmentation and individuation of hypercycles, giving in effect protocells, might be expected to result in a selective advantage (Eigen, 1971):

(a) The catalytic hypercycle presupposes a coherent interaction between numerous molecular components (see Section 11.1). This complex space- and time-dependent structure can more easily be preserved if compartmentation prevents the system from diffusing apart.

(b) Only compartmented hypercycles can exploit functionally advantageous branches for their own exclusive advantage (*cf.* Section 9.3). Similarly, disadvantageous branches (parasites) can be kept under control more effectively in compartmented hypercycles, so that the danger of a lethal infection can be kept at bay.

(c) A compartmented hypercycle can, once it has attained a sufficiently high degree of accuracy, take on an individual nature. Individuation takes place when a hypercycle unites its entire information content, which was previously distributed among several information carriers, into a single, stable chain which is thereafter reproduced as an individual genome. The process of individuation converts the hypercycle to a genuinely "self-reproducing" system and gives it the status of a protocell. If the connection of the information units takes place *e.g.* by means of a ligating enzyme (ligase) then the resulting genome - as in the bacteria - is likely to be circular.

Compartmentation and individuation abolish the non-linear selection characteristic of hypercycles, and the convergent phase of hypercyclic evolution goes over into the divergent phase of Darwinian evolution, which has led to the great variety of species of today.

What physical boundary conditions could lead to the formation of compartments? The following possibilities are conceivable:

(1) NATURAL COMPARTMENTATION. This is based upon inhomogeneities in the structure of the environment. The primordial earth must have offered a multitude of such natural inhomogeneities. Possible types include grains of sand, surfaces of minerals, and so on. Some processes of self-organization at porous surfaces have been described by H. Kuhn (1976).

(2) PHYSICAL AUTOCOMPARTMENTATION. Spatial patterns can arise in solutions purely on the basis of physical interactions between the molecular components (coacervate drops, polyelectrolyte latticing *etc.*). Coacervates, first described by A. Oparin (1965), or microspheres, which consist mainly of lipids or protenoids with hydrophobic side chains (Fox, 1965), form spontaneously under prebiotic conditions. They can be prepared in the test tube just like, for example, lipid vesicles.

(3) CHEMICAL AUTOCOMPARTMENTATION. Specific catalytic mechanisms can lead to spatially ordered patterns, termed *dissipative structures* (Nicolis and Prigogine, 1977).

(4) SYSTEM-INDUCED COMPARTMENTATION. It is conceivable that advanced hypercycles could synthesize special substances, such as ingredients of membranes, which lead to physical or chemical autocompartmentation.

To summarize: Compartmentation and individuation *do not* follow directly from principles of hypercyclical organization. Neither can they replace the information-integrating properties of hypercycles (see the discussion of Eigen *et al.*, 1980). Compartmentation and individuation *can* result in a higher survival probability for hypercycles and offer in this way an evolutionary advantage. Seen thus, the formation of protocells appears as an inevitable last stage in the process of molecular self-organization.

IV. Model and Reality

12. Systems Under Idealized Boundary Conditions

The evolution of life is a process in which the structure of the historical boundary conditions played an important rôle. Physics, in contrast, seeks to utilize suitable abstractions of the boundary conditions in order to make predictions, of the greatest possible generality, about the physical structure of real systems. A physical theory is only useful when it leads at least in principle to quantitative and reproducible experiments.

A physical theory of evolution must therefore involve suitable abstractions of the historical boundary conditions, if it is to make predictions on a level of generality higher than that of the specific system. In this chapter we therefore examine the thermodynamic conditions of evolution in general and some boundary conditions germane to our theory in particular.

12.1. The Thermodynamic Framework

A catalyst is defined as a substance which accelerates a chemical reaction without itself being changed in the course of the reaction. Thus a catalyst changes the rate at which chemical equilibrium is reached without changing the position of equilibrium. The law of microscopic reversibility requires further that a catalyst influence the forward and backward reactions equally effectively. For this reason, autocatalysis in systems at or near equilibrium cannot lead to the drastic changes in concentration which are needed for efficient selection; selective self-organization can only occur in systems far from equilibrium, where chemical reactions take place irreversibly.

We do not intend here to lay the thermodynamic foundations of evolution theory; this topic has been the subject of fundamental investigations by I. Prigogine and his colleagues, and it is summarized in the book "Self-Organization in Nonequilibrium Systems" by G. Nicolis and I. Prigogine (1977). What we wish to show are the points of contact between the molecular theory of evolution developed here and the thermodynamic theory of irreversible processes. Before formulating the thermodynamic conditions for molecular self-organization, we must first examine the system constraints needed for irreversible processes to occur. Classical thermodynamics tells us that *isolated* systems (*i.e.*, systems which cannot exchange energy or matter with their surroundings)* always proceed towards equilibrium. During the attainment of equilibrium the entropy (S) of the system increases monotonically and reaches a maximum in the equilibrated state:

$$\frac{dS}{dt} \geq 0 \tag{12.1}$$

This is the second law of thermodynamics. It has two consequences relevant to our discussion.

(1) A process leading towards equilibrium ($dS/dt > 0$) is *irreversible* since a change of direction would violate the second law. A change for which $dS/dt=0$ is however *reversible*.

(2) All deviations from equilibrium lead to a decrease in entropy. A system in equilibrium therefore reacts to fluctuations by reverting to equilibrium (Fig. 12.1); in other words, fluctuations around the equilibrium state are self-regulating. This "control mechanism", inherent to all equilibrium systems, has already been discussed in connection with the Ehrenfest game (compare Figs. 6.2 and 12.1). Since all evolutionary changes are ultimately due to the macroscopic reinforcement of individual fluctuations, it is clear that selection and evolution will never be observed in systems at equilibrium.

We shall therefore examine systems which are well away from equilibrium. In order to prevent these systems from collapsing into an equilibrium state, they must be

---

**Cf.* a *closed* system, which can exchange energy but not matter and an *open* system which can exchange both.

continually supplied with free energy or with energy-rich matter. Such open systems can have time-independent states in which the concentrations of the molecular components do not change with time. These *stationary states* or *flow equilibria* correspond exactly to our concept of selection equilibrium defined in Section 5.2.

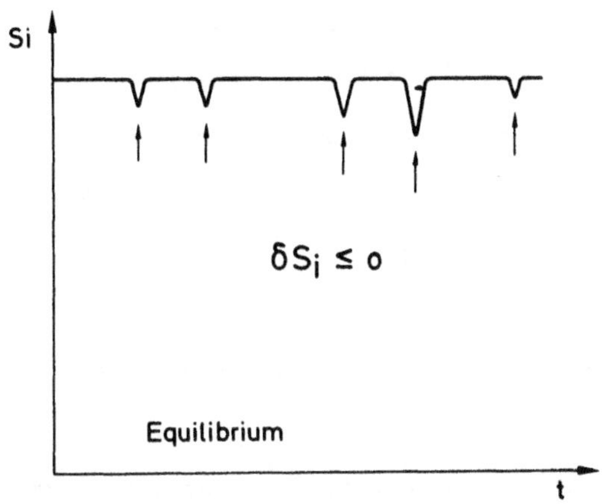

Fig. 12.1. Entropy fluctuations in stable equilibrium.

In this section we shall look for a general thermodynamic criterion of stability for stationary states in CP systems.

The change in entropy in an open or a closed system is made up of two contributions:

$$dS = d_iS + d_eS \tag{12.2}$$

Here d_iS is the entropy produced within the system and d_eS the entropy entering the system through its boundaries, *e.g.* by transfer of heat (in open and closed systems) or of matter (in open systems only). The second law of thermodynamics implies that the entropy production within the system is greater than or equal to zero:

$$d_iS \geq 0 \tag{12.3}$$

We must now distinguish between reversible and irreversible processes. Only irreversible processes, such as chemical reaction, the conduction of heat *etc.* contribute to the production of entropy, since for reversible processes we have:

$$d_iS = 0 \tag{12.4}$$

It is difficult in general to calculate the amount of entropy produced by an irreversible process. We shall take as an example the increase in entropy in a *closed* system in which a chemical reaction written in the general form

$$\nu_1 A_1 + \nu_2 A_2 + \cdots\cdots \rightarrow \nu_1' A_1' + \nu_2' A_2' + \cdots\cdots \tag{12.5}$$

is taking place at constant temperature and pressure. (Note that by convention the ν_i values (reactants) are negative.) We can easily calculate the production of entropy if we assume that away from equilibrium the Gibbs formula for entropy

$$dS = d_eS + d_iS = \frac{1}{T}\,dU + \frac{P}{T}\,dV - \sum_\gamma \frac{\mu_\gamma}{T}\,dn_\gamma \tag{12.6}$$

still holds, where U is the internal energy, V the volume, n_γ the number of moles of the γth component and μ_γ the molar Gibbs free energy or *chemical potential* of component γ. The validity of Eq. (12.6) can only be proven close to equilibrium.

In order to continue with our derivation of a stability criterion for stationary states, we now introduce some definitions. The equation

$$d\xi = \frac{1}{\nu_\gamma}\,dn_\gamma \qquad \text{(for all } \gamma\text{)} \tag{12.7}$$

defines the *extent* ξ of a chemical reaction, so that the quantity

$$v = \frac{d\xi}{dt} \tag{12.8}$$

represents the rate of reaction. The *affinity* A of a chemical reaction is related to the chemical potentials μ_γ by the definition

$$A = - \sum_\gamma \nu_\gamma \mu_\gamma \tag{12.9}$$

Using Eqs. (12.7) and (12.9) we can rewrite the Gibbs formula:

$$dS = d_e S + d_i S = \frac{1}{T}\, dU + \frac{P}{T}\, dV + \frac{1}{T}\, A\, d\xi \tag{12.10}$$

For a closed system, it follows from the first law of thermodynamics that the entropy influx is given by

$$d_e S = \frac{1}{T}\, dU + \frac{P}{T}\, dV = 0 \tag{12.11}$$

so that we obtain from Eqs. (12.10) and (12.11)

$$d_i S = \frac{1}{T}\, A\, d\xi \geq 0 \tag{12.12}$$

in which the inequality follows from Eq. (12.3). We now apply the definition in Eq. (12.8) and obtain the explicit expression for the rate of entropy production σ:

$$\sigma = \frac{d_i S}{dt} = \frac{A}{T} \cdot v \geq 0 \tag{12.13}$$

The entropy production rate is thus the reaction flux v multiplied by the *reaction force*, A/T.

The thermodynamic theory of irreversible processes shows that relation (12.13) is only a special case of a general law, which lays down that (near equilibrium) the internal entropy production σ can be represented as a bilinear form of generalized fluxes J_i and forces X_i:

$$\sigma = \sum_i J_i X_i \tag{12.14}$$

Irreversible process	Generalized force	Generalized flux
Chemical reaction	$A = -\sum_{\gamma} \nu_{\gamma} \mu_{\gamma}$ (A = Affinity)	Rate of reaction per unit volume
Heat conduction	$-T^{-2}$ grad T (T = Temperature)	Rate of energy flow per unit area
Binary diffusion	- grad μ_{γ} (μ_{γ} = chemical potential)	Rate of flow of matter per unit area
Electrical conductance of binary electrolyte	- grad ϕ (ϕ = electrostatic potential)	Current density

Table 12.1. Irreversible processes and the forces and fluxed connected with them. (After Moore, 1972.)

Examples of such fluxes and forces are given in Table 12.1. Equation (12.14) is the fundamental formula of the macroscopic thermodynamics of irreversible processes. It applies to all irreversible changes in open, closed and isolated systems close to equilibrium.

We shall now examine the behaviour of σ when the stationary state is approached. In order to do so we need one further relationship.

It is obvious that for all conceivable processes in a system at thermodynamic equilibrium

$$J_i = 0 \quad \text{and} \quad X_i = 0 \tag{12.15}$$

It is therefore reasonable to assume linear, homogeneous relationships between forces and fluxes in systems close to equilibrium, *i.e.*,

$$J_i = \sum_j L_{ij}X_j \qquad (12.16)$$

The region of validity of Eq. (12.16) defines the scope of the *linear thermodynamics of irreversible processes*. In certain cases (*e.g.* the flow of heat) the linear proportionality between the flux and force terms extends over a wide range. In other cases, such as chemical reactions, Eq. (12.16) applies only for small deviations from equilibrium.

The quantities L_{ij} are called *phenomenological* coefficients. L. Onsager (1931a,b) made the important discovery that for the coefficients L_{ij} the *reciprocity relations*

$$L_{ij} = L_{ji} \qquad (12.17)$$

apply in the linear region.

The Onsager relations (12.17) have been experimentally verified many times over. They contained the first indication that there exist general statements about non-equilibrium processes which are independent of particular molecular pathways.

Another important theorem of linear thermodynamics is the stability criterion for stationary states (Prigogine, 1945; Glansdorff and Prigogine, 1971). The Prigogine-Glansdorff principle has a certain relevance to the molecular theory of evolution, as we shall see below.

We consider a system in which a constant temperature difference is maintained between two surfaces, and allow energy and matter to be transported between the two surfaces. However, no matter can enter or leave the system.

There are then two forces, X_e and X_m, defined respectively by the differences in temperature and concentration and two fluxes, that of energy (J_e) and that of matter (J_m).

After a sufficiently long time the system will have reached a state in which redistribution of matter will have removed the difference in chemical potential, while the transport of energy between the boundary surfaces will be in a stationary state.

As we have seen above, entropy increases continually in a non-equilibrium stationary state. In the linear region (*i.e.* close to equilibrium), we have

$$\sigma = J_e X_e + J_m X_m > 0 \tag{12.18}$$

and, since the phenomenological equations (12.16) apply,

$$J_e = L_{ee} X_e + L_{em} X_m \tag{12.19}$$

and

$$J_m = L_{me} X_e + L_{mm} X_m \tag{12.20}$$

We now substitute Eqs. (12.19) and (12.20) into Eq. (12.18) and, combining the result with Eq. (12.17), obtain for the rate of entropy production the expression

$$\sigma = L_{ee} X_e^2 + L_{me} X_e X_m + L_{mm} X_m^2 > 0 \tag{12.21}$$

We then differentiate Eq. (12.21) with respect to X_m at constant X_e and obtain the partial derivative

$$\frac{\partial}{\partial X_m} (\sigma) = 2(L_{me} X_e + L_{mm} X_m) = 2\, J_m \tag{12.22}$$

The system-imposed constraint of zero material flux (J_m=0) leads to the extremum principle

$$\frac{\partial}{\partial X_m} (\sigma) = 0 \tag{12.23}$$

Equation (12.23) can be generalized. However, it applies only in systems close to equilibrium, since it assumes that the phenomenological equations are linear, the

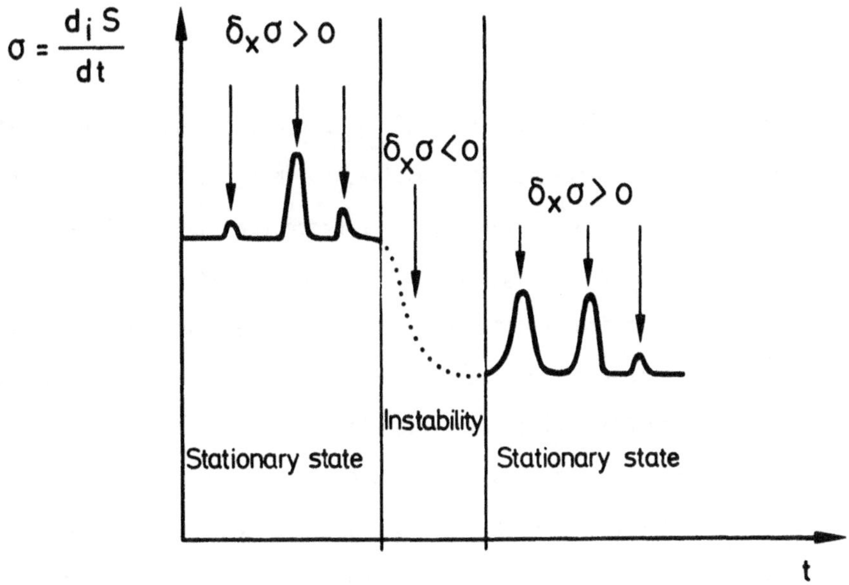

Fig. 12.2. Fluctuations in the internal entropy production of a system in a stationary state close to equilibrium.

coefficients L_{ij} are constant and the reciprocity relations (12.17) are valid. The extremum principle (12.23) expresses what I. Prigogine (1979) has called the "inertia" of non-equilibrium systems: as long as the boundary conditions prevent the system from attaining thermodynamic equilibrium (a state of zero entropy production) then the system moves towards a state of "minimal dissipation".

The Prigogine-Glansdorff principle states that *a stationary state running at constant fluxes is stable with respect to a fluctuation if and only if the fluctuation involves a non-negative change in the rate of entropy production.* In other words, the stationary state will not collapse as long as the condition

$$\delta_X \sigma \geq 0 \tag{12.24}$$

is satisfied. In contrast to genuine equilibria, stationary states are *metastable*. Whenever a fluctuation appears for which

$$\delta_X \sigma < 0 \qquad (12.25)$$

the system becomes unstable (see Fig. 12.2) and moves towards a new, metastable state. The change in S_i is necessarily negative, so that the new stationary state has a lower entropy and thus a greater degree of order than the previous one*.

It has been pointed out by M. Eigen (1971) that in autocatalytic reactions, such as the self-instructed replication of nucleic acids, the appearance of mutants with a selective advantage may lead to just such negative fluctuations in the production of entropy. The Prigogine-Glansdorff principle states, in effect, that non-equilibrium can generate ordered states. It thus refutes the frequently-raised objection that the increase in order associated with the evolution of living systems is incompatible with the increase in disorder implied by the second law. *The increase in complexity concomitant with evolution is in full accordance with the laws of physics.*

However this does not mean that the molecular theory of evolution provides a thermodynamic explanation of evolution based upon the Prigogine-Glansdorff principle, since the generality of this principle is restricted to systems (i) in stationary states, (ii) at constant flux and (iii) near equilibrium. At no point have we made use of the Prigogine-Glansdorff principle in developing the molecular theory of evolution. The best we can say at present is that our theory is congruent with that of the thermodynamics of open systems.

---

*The validity of the Prigogine-Glansdorff criterion has been the subject of critical discussions (*cf.* Landauer, 1975). It is obvious that the entropy production principle does not describe what happens at the instability point; neither does it determine the resulting structures or provide a classification for them.

12.2. Stationary and Periodic Boundary Conditions

In developing our theory we have considered rather simple self-organizing systems, and in particular ones which satisfy CP or CF constraints. The use of time-independent boundary conditions does, of course, not mean that the actual, historical process of evolution of macromolecules took place under such idealized conditions. On the contrary, it is more plausible that the historical boundary conditions involved an undefined and time-dependent combination of CP and CF constraints. However, our theory should not for this reason be regarded as irrelevant. A parallel can be taken from the history of physics: the abstract Carnot cycle, originally intended as a method of calculating the maximum efficiency of a steam engine, made an important contribution to our understanding of equilibrium thermodynamics, even though a steam engine could never run under equilibrium conditions.

In a similar way, CP and CF boundary conditions provide a framework within which the molecular theory of evolution can be described mathematically in fairly simple terms and which none the less allows conclusions about the principles of molecular self-organization. In addition to this the CP and CF constraints provide a basis for the design of quantitative evolution experiments in the laboratory (see Chapter 13).

To recapitulate: the assumption of steady-state conditions is by no means an indispensible prerequisite for our theory, although it is of great practical value. The fact that the principal results of this theory do not depend upon the special choice of boundary conditions has been illustrated in Chapter 8, where we deliberately dispensed with the CP condition and were still able to reach general conclusions about various processes of self-organization.

Ways in which the boundary conditions may be fitted to historical reality have been investigated in a series of papers by H. Kuhn (1972, 1976 and 1977) and H. Kuhn and J. Waser (1981). These authors, too, set out from more or less idealized boundary conditions. However, they presuppose not a homogeneous and stationary environment but a periodically structured one, varying with night and day, shadow and light, *etc.* Such a periodicity in time and heterogeneity in space lead automatically to periodicity in temperature and composition, which could have been a source of synchronized enzyme-free replication of biological macromolecules. However the attribution of overriding importance to special boundary conditions raises a number of difficulties and does not solve the basic problems connected with the origin of life. There is no way in which a periodically-structured environment can provide the *sole* impetus for overcoming the information threshold (Chapter 7) and thus allowing the unrestricted evolution of complex structures. As long as the prevailing material conditions allow the initiation of a primary process of biological self-organization, there is no reason to formulate our theory in terms of especially complex boundary conditions.

The assumption of steady (and thus regulable) environmental constraints allows experimentally reproducible boundary conditions to be set up (*e.g.* by flux regulation). This represents a great advantage of molecular theory in comparison with the models of population genetics. In the latter, the boundary conditions cannot be varied: they must be inserted into the theory just as they are found.

13. Evolution in the Test-Tube

Experiments have shown beyond reasonable doubt that under the conditions prevailing on the primordial earth, that is, in a reducing atmosphere supplied with various forms of energy, both nucleic acids and proteins could form spontaneously and independently of one another.* Thus the mere existence of biological macromolecules is sufficiently explained by the known laws of physics and chemistry. For this reason we have formulated the question of the origin of life in terms not of the chemical origin of nucleic acids and proteins, but rather of the functional coupling between them which might lead to the assembly of a self-reproducing and evolving genetic unit.

The theory presented here is based on the idea that selection and evolution in the Darwinian sense must underlie any molecular self-organization process if it is to lead to the nucleation of a living system. Any kind of experimental verification of this theory will therefore depend on how well patterns of natural selection can be seen among nucleic acids and proteins in non-living systems.

We shall describe in the sections to follow the present state of experimental work aimed at answering this question. At the focal point of experimental analysis is the RNA replication system of the virus Q_β first isolated by S. Spiegelman and his co-workers (Haruna and Spiegelman, 1965) and used for the cell-free synthesis of biologically active nucleic acids. The Q_β system is described in Section 13.1; the amount of detail is reduced to that necessary for an understanding of the selection and evolution experiments, which follow in Section 13.2. Finally, Section 13.3 introduces an experimental design intended to allow the quantitative assessment of evolution experiments in terms of the theory set out so far.

---

*There is a considerable literature on problems of prebiotic chemistry, of which the monograph by S. Miller and L.E. Orgel (1974) may be recommended for an introductory study.

13.1. The Q_β Replicase System

The simplest living beings are the viruses. They are so simple that they do not even possess their own metabolic system, and certain viruses (the *viroids*) consist solely of an unprotected genome. Although (or because) the viruses are clearly at the threshold of life (*cf.* Chapter 1), they have proved ideal test subjects for studying the basic processes of living systems.

In the following treatment we shall be concerned with the virus Q_β. This is one of the RNA-containing viruses which infect the bacterium *Escherichia coli*, and to this class of viruses (called *bacteriophages* or, for short, *phages*) belong also the viruses f2, R17 and MS2.

We start by considering a problem which confronts all RNA viruses: In accordance with the "central dogma" of molecular biology all of the host cell's RNA molecules are synthesized on a DNA template. This means, if the synthesis of viral RNA is also instructed by DNA, that either the host cell DNA must contain a sequence homologous to that of the viral RNA or such a DNA sequence must be produced, after infection, by *reverse transcription*, from RNA to DNA. However, no homology has ever been found between viral RNA and the DNA of infected cells, which led I. Haruna *et al.* (1963) to the conclusion that the RNA phages must have developed a reproductive mechanism independent of DNA. This was confirmed by the discovery of an RNA-dependent RNA polymerase (*replicase*) whose synthesis is induced after a cell has been infected with a viral genome, and by means of which the viral RNA is copied directly.

None the less, it is clear that the mere existence of an RNA polymerase does not completely solve the problem of viral self-reproduction. It must be remembered that the viral genome operates in a highly "heterogenetic" environment, *i.e.*, one which carries a very large number of "alien" mRNA molecules, those of the host cell. The virus therefore needs a further mechanism to ensure that its own replicase reproduces the viral RNA in preference to the host RNA. This selectivity is the more important since any one RNA molecule has only a short life-span in the living cell.

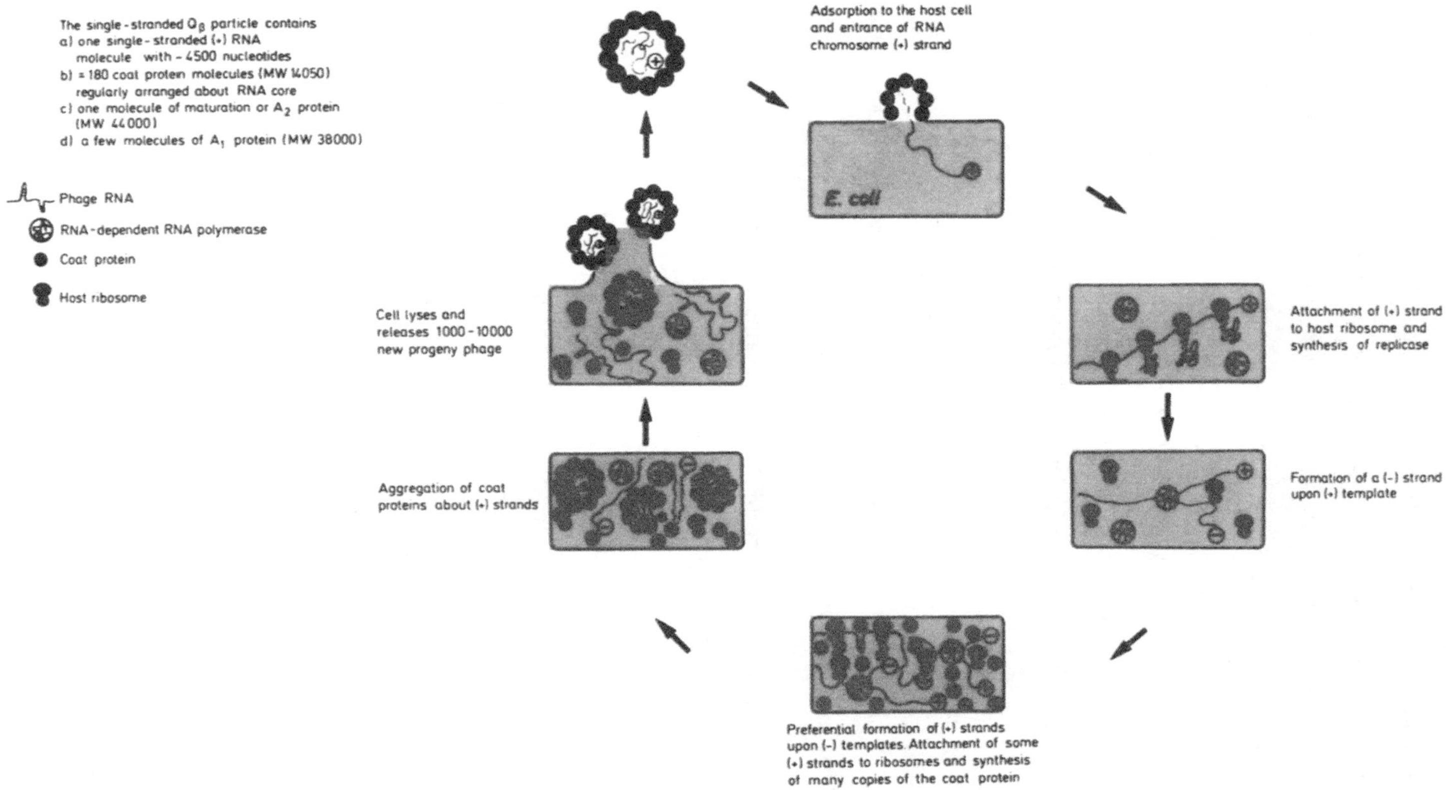

Fig. 13.1. The life cycle of the single-stranded RNA virus Q_β. (Based on Watson, 1976.)

S. Spiegelman, pursuing this argument, suggested that the replicase whose existence he had postulated must be able to discriminate between viral and non-viral mRNA, so that it could copy viral RNA efficiently and be inactive in the replication of other RNA molecules. With his co-workers Spiegelman was able to confirm this presumption experimentally by isolating a template-specific, RNA-dependent RNA polymerase from phage-infected *E. coli* cells (Haruna *et al.*, 1963; Haruna and Spiegelman, 1965).

Let us take a closer look at the life cycle of the Q_β phage (Fig. 13.1). With the help of its specific replicase, a single Q_β phage genome produces around 1000 to 10 000 progeny. The single-stranded viral genome fulfils three functions:

(1) It forms an integral part of the infectious phage particle.

(2) After the host cell has been infected, it acts as messenger RNA for the synthesis of phage-specific proteins.

(3) It is recognized by one of these proteins, the phage-induced replicase, and is then itself replicated preferentially.

The genome of phage Q_β is made up of about 4500 nucleotides and possesses one *cistron* for each of the following proteins: maturation protein (A_2 protein), coat protein and the β subunit of the phage-induced replicase (Fig. 13.2a). Alongside these three proteins there occurs a fourth virus-specific protein, referred to as A_1 protein. This is thought to be a product of occasional failure of the host ribosomes to recognise the "stop" codon UGA at the end of the cistron coding for the coat protein, resulting in a longer protein being synthesized. Both A_1 and A_2 proteins are found in the Q_β virus particle and both seem to be essential for its ability to infect.

The Q_β replicase has been characterized biochemically in detail (Fig. 13.2b). The *holoenzyme* has a molecular weight of 215 000 and is composed of four subunits. Only the β subunit is encoded in the phage RNA. The other three subunits are polypeptides appropriated from the host cell and which normally have a completely different function, in the cell's protein-synthetic apparatus.

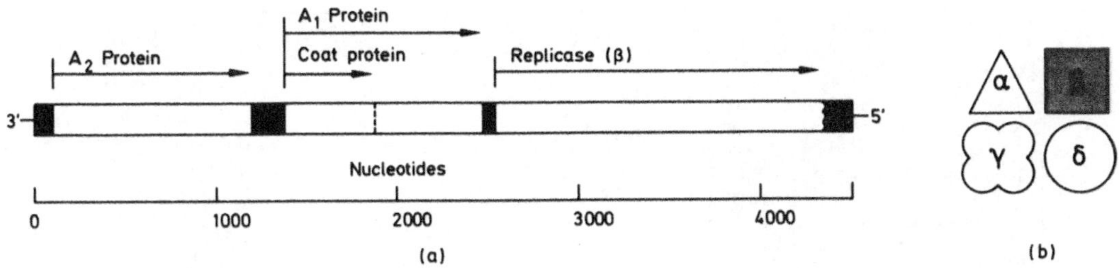

Fig. 13.2. Schematic representation of the genetic organization of the bacteriophage Qβ. (a) Arrangement of the four genes. At both ends of the genome and between all consecutive cistrons are RNA segments which are not translated into protein. These regions are indicated in black. (b) Subunit structure of Qβ replicase. The holoenzyme is made up of one polypeptide chain, coded for by the virus, and three that are taken from the host: α is the ribosomal protein S1, β is the subunit for which the virus codes, γ is the elongation factor Tu of protein synthesis and δ is the elongation factor Ts of protein synthesis.

As regards the functions of the individual subunits in Q_β replicase, it is currently believed that: (a) The α subunit is necessary for the binding of the replicase to the (+) but not to the (-) strand of phage RNA. (b) The β subunit is the polymerizing factor in the replicase. (c) The γ and δ subunits are (like the α subunit) involved in the initiation step of RNA replication.

The RNA replication system of Q_β phage can be isolated and used for the cell-free propagation of Q_β RNA. The reaction mixture (Q_β replicase system) required for the replication of Q_β RNA *in vitro* is given in Fig. 13.3. This system is extremely convenient for studying the basic processes of the enzymatic replication of nucleic acids.

Experimental results to date have led to a comprehensive picture of phage RNA replication *in vitro*. Not surprisingly, the reproduction of viral RNA follows the cross-catalytic reproduction principle with which we are already familiar (*cf.* Fig. 2.4). Without going into complex details of the reaction mechanism, which has been investigated by C. Biebricher *et al.* (1981a,b), we can distinguish two steps in the overall process:

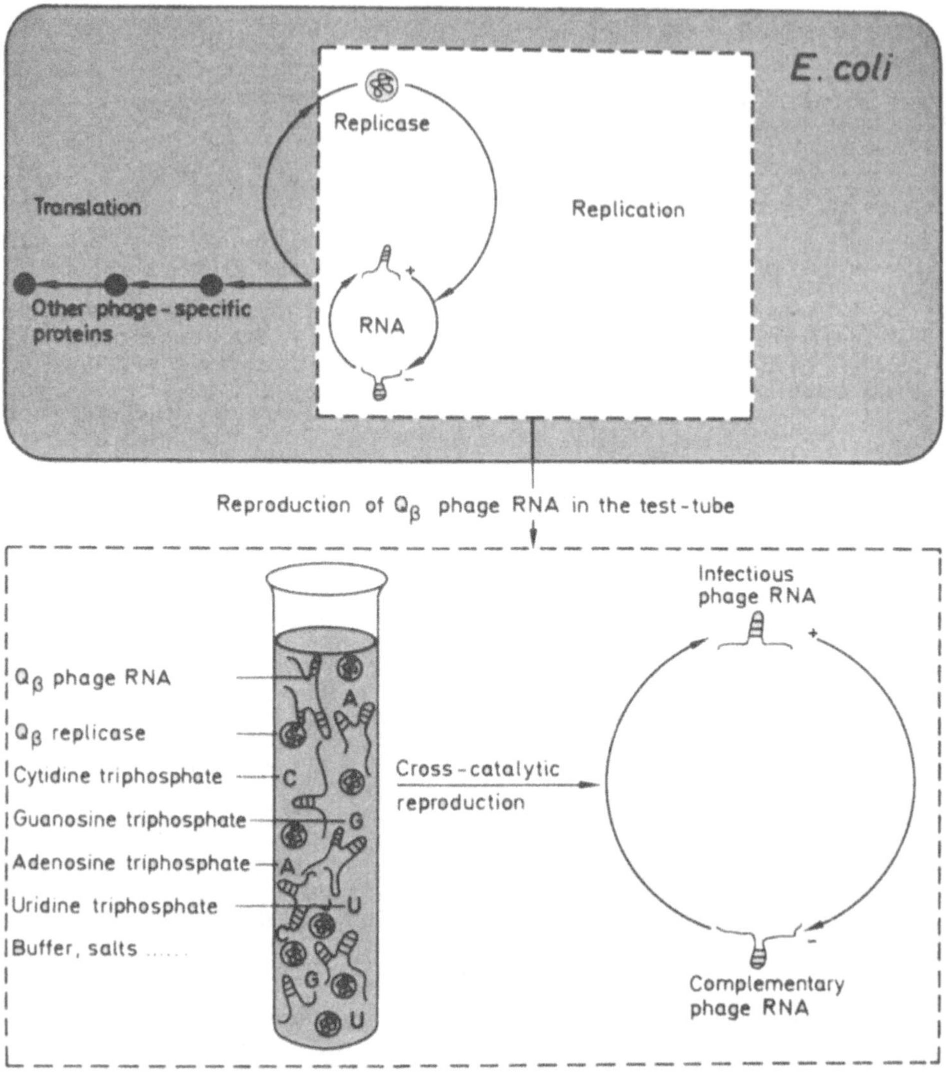

Fig. 13.3. Reproduction scheme of Q_β phage RNA. The cross-catalytic reproduction of the genetic information unit I_+ (*cf.* Fig. 11.1b) of the phage takes place by complementary base recognition. This reproduction cycle is coupled hypercyclically within itself by way of a replication enzyme E. The coupling involves: (1) Translation of the phage genome I by the protein-synthetic apparatus of the host cell (the latter may be regarded as a constant environmental factor) and (2) the reproduction of the viral RNA by the template-specific replicase. In addition to the coupling factor E the phage genome contains the information for three further proteins which are required for its reproduction *in vivo* (see Fig. 13.2).

The RNA replication system of the bacteriophage Q_β can be isolated and made to work outside the bacterial host cell.

(1) The replicase binds the 3′ end of the (+) strand and then uses it as a template for the synthesis of a (-) strand. A template can bind only one replicase molecule in the initiation phase, but as the replicase molecule moves away from the 3′ end it can be followed by another, and so on. The (-) strands are synthesized in the direction 5′ to 3′.

(2) When each replicase molecule has completed its (-) strand, finishing it at the 3′ end, it can use the product as a template for a new (+) strand, which is therefore likewise synthesized in the direction 5′ to 3′. The replicative intermediate thus always consists of a single-stranded (+) or (-) template, on which one or more replicase enzymes are each synthesizing a complementary copy.

It is of interest that both (+) and (-) strands emerge from the synthesis as single-stranded products; it is not yet clear how the replicative intermediates are prevented from associating to give double-stranded RNA, which the complementarity of template and product would lead one to expect.

In summary, the Q_β replicase system possesses the prerequisites sufficient to induce selection and evolution among biological macromolecules (*cf.* Section 5.1):

(1) The system manifests *metabolism*, in that the replication of RNA is concurrent with the turnover of energy-rich ribonucleoside triphosphates into energy-deficient ribonucleoside monophosphates.

(2) The system shows *self-reproduction*, since it includes the autocatalytic formation of new Q_β RNA molecules.

(3) The system is prone to *mutation*, since the synthesizing machinery (Q_β replicase) possesses only a limited accuracy (*cf.* Table 7.3).

In recent years, the Q_β replicase system has been used in a variety of evolution experiments, whose design and outcome fit very well into the theory which we have outlined. A detailed review of such results has been published elsewhere (Küppers, 1979). We shall discuss below some selected experiments.

13.2. Darwinian Selection in vitro

The division of biological macromolecules into genotype (DNA) and phenotype (protein), as found today in the living cell, could only have become possible once a genetic code had developed. During the earliest phase of evolution, before a translation apparatus existed, the phenotypic expression of genetic information must have lain in the phenotypic properties of the genetic material itself. It is therefore probable that the first storage unit of biological information was the RNA molecule, since, in contrast to DNA, RNA can by virtue of its chemical structure adopt complex tertiary structures and thus manifest specific phenotypic behaviour (see Fig. 13.4).

In order to simulate the earliest phase of molecular evolution, experiments must be designed in such a way as to allow selection pressure to operate directly on the genetic material and not by way of the gene product. As will be seen, the Q_β replicase system is especially suited for such experiments.

The first evolution experiments with a self-replicating RNA molecule *in vitro* were carried out by Spiegelman and his co-workers. In these experiments selection pressure was induced by stepwise dilution, using a *serial transfer* technique, illustrated in Fig. 13.5. The prototype serial transfer experiment was carried out with a standardized reaction mixture, which contained all the ingredients needed for the optimal replication of Q_β phage RNA *in vitro* (for details see Figs. 13.3 and 13.5). At the beginning of the experiment the reaction was started by the addition of RNA from Q_β. The RNA molecules were then allowed to replicate for a certain time interval, at the end of which a portion of the reaction solution was withdrawn and added to fresh, RNA-free medium. The reaction mixture was then subjected to further cycles of incubation followed each time by dilution. By reducing the incubation time between successive dilution steps, the RNA population was kept roughly constant even though its average excess productivity rose continuously during the experiment. In a typical experiment the incubation time was reduced from 20 minutes (transfers 1-13) to 15 minutes (transfers 14-29), then to 10 minutes (transfers 30-38), 7 minutes

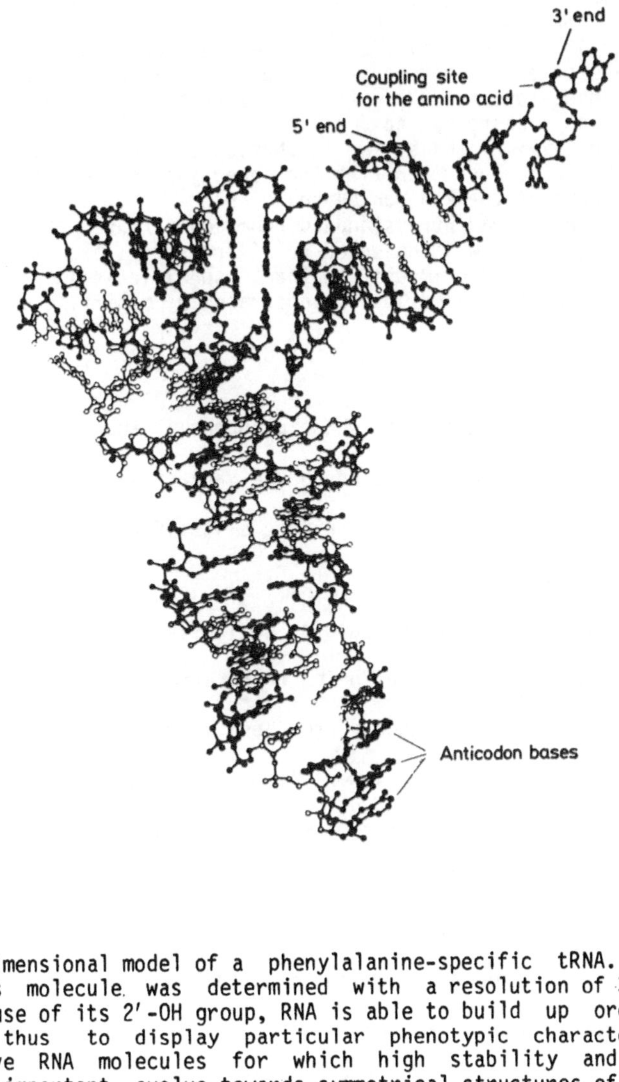

Fig. 13.4. Three-dimensional model of a phenylalanine-specific tRNA. The spatial structure of this molecule was determined with a resolution of 3 Å using X-ray diffraction. Because of its 2'-OH group, RNA is able to build up ordered tertiary structures and thus to display particular phenotypic characteristics. All functionally active RNA molecules for which high stability and preferential reproduction are important evolve towards symmetrical structures of the kind shown here (*cf.* Figs. 10.3 and 11.3). (From Rich and Kim, 1978.)

(transfers 39-52) and finally to 5 minutes (transfers 53-74). The experiment was stopped after the 74th transfer and the reaction products were analysed.

The progress of a typical serial transfer experiment is shown in Fig. 13.6. After a certain number of transfers the RNA molecules multiply considerably more rapidly than at the beginning of the experiment. At the same time, however, their ability to infect bacteria has been lost.

Fig. 13.5. Scheme of a serial transfer experiment. With a special dilution technique, selection pressure can be applied to the Q_β replicase system. The RNA component adapts itself, by evolution, to the reaction conditions chosen.

A serial transfer experiment is started by inoculating a reaction mixture (tube 1) at time t=0, with Q_β-specific RNA. After a certain incubation time an aliquot of the reaction mixture is transferred to fresh medium (tube 2). The dilution procedure is then repeated n times, while the incubation time between the individual dilution steps is made shorter and shorter. If the energy-rich monomers are radioactively labelled, the growth of the RNA can be followed.

In the prototype experiment a standard reaction mixture was used. The starting material, Q_β phage RNA, was subjected to 74 transfers, each involving dilution by a factor 12.5. In later experiments further, specific selection constraints were superimposed upon the general dilution constraint by modifying the reaction mixture. ▶

Detailed analysis of the experiment whose progress is shown in Fig. 13.5 showed that the selection pressure had produced an RNA variant which possessed only 17% of the original genetic material and which reproduced itself fifteen times more rapidly than the original, complete phage RNA.

The dilution accompanying each transfer obviously gave those molecules that could grow rapidly the greatest chance of being transferred into fresh growth medium. Each mutation in the nucleotide sequence of the "evolving" phage RNA leading to a higher rate of replication or to a lower rate of degradation meant an advantage (*cf.* Eq. 5.17) and the RNA bearing it became selected. This selective advantage might be associated with (i) the initiation step (binding of Q_β replicase to the RNA template), (ii) the elongation steps (incorporation of nucleotides into the growing RNA copy) or (iii) the termination step (dissociation of the replicase from its template after completion of the copy).

In the experiment of Fig. 13.6 the increased rate of RNA synthesis is clearly due in part to the shortening of the nucleotide sequence of the "evolving" RNA molecules. As long as the region needed for recognition of the replication enzyme is unaffected by the shortening, maintenance of the selection pressure should lead to shorter and shorter RNA variants. This explanation is supported by serial transfer experiments carried out by R. Saffhill *et al.* (1970).

The original interpretation of the results from the serial transfer experiment was later called into question on the basis of recent investigations of the Q_β system. M. Sumper and R. Luce (1975) showed that highly purified Q_β replicase can also build

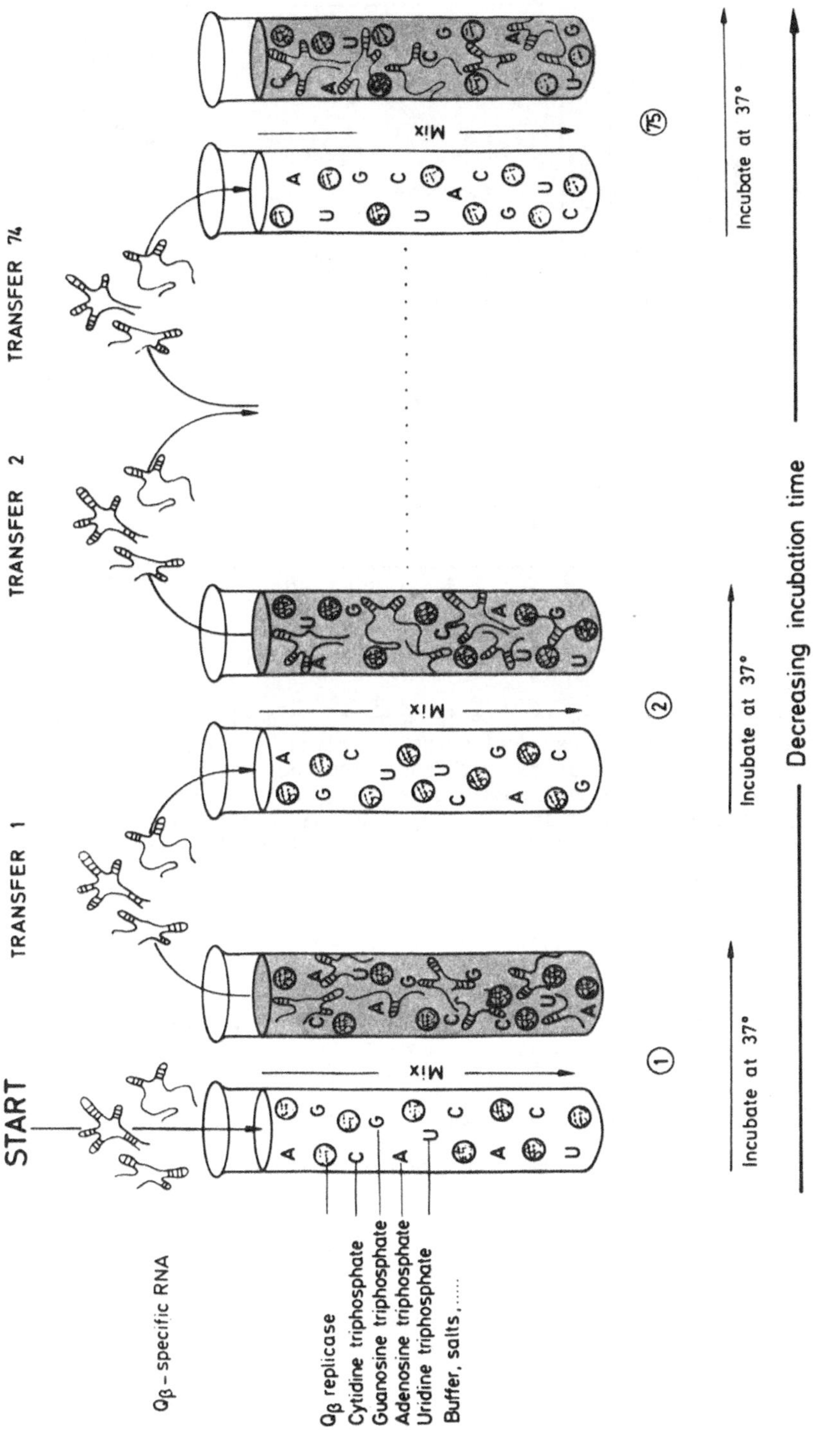
START
TRANSFER 1
TRANSFER 2
TRANSFER 74
Qβ - specific RNA
Qβ replicase
Cytidine triphosphate
Guanosine triphosphate
Adenosine triphosphate
Uridine triphosphate
Buffer, salts, ...
Mix
1
2
75
Incubate at 37°
Incubate at 37°
Incubate at 37°
Decreasing incubation time

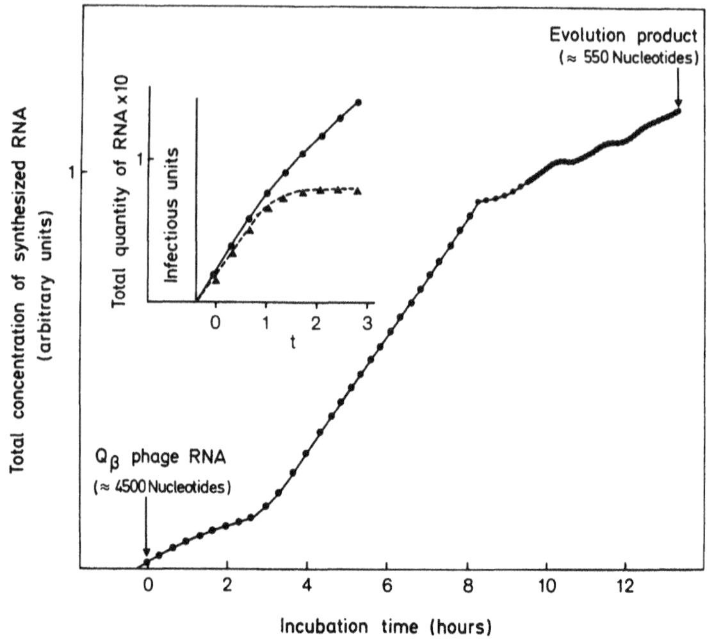

Fig. 13.6. Evolution diagram of a serial transfer experiment with Q_β phage RNA. The total amount of RNA synthesized during the experiment is plotted along the ordinate. At various times an aliquot was withdrawn from the reaction solution and a product analysis (test of infectiousness, molecular weight determination *etc.*) was carried out. Inset: the infectiousness of the Q_β phage was rapidly lost, since there was no selection pressure to maintain this property. (After Mills *et al.*, 1967.)

up small RNA molecules from energy-rich monomers without instruction by an exogenous RNA template. If such a product *de novo* happens to have the correct recognition signal for Q_β replicase, then it will act as an endogenous template and be amplified selectively. Moreover, it was found that the resulting RNA structures were always adapted to the reaction conditions under which they nucleated.

It is therefore not possible, on the basis of the experimental data available, to exclude the alternative explanation that during the serial transfer experiment shown in Fig. 13.6 an RNA synthesis has taken place *de novo* and that its products have interfered with the "evolution" of the primary phage RNA. The question of the mechanism of this evolution can only receive a satisfactory answer when all the stages through which the RNA molecules evolve are known.

With this aim in mind, D.R. Mills *et al.* (1973) isolated and sequenced an RNA which had been selected under defined, standard conditions in a reaction *de novo* (Fig. 13.7). They christened this molecule, which in contrast to Q_β phage RNA possesses a chain length of only 218 nucleotides, *midivariant* RNA (MDV-RNA).

MDV-RNA has a conspicuously high degree of intramolecular complementarity and thus a marked secondary structure. Further, it is highly symmetrical. This is characteristic of self-replicating RNA structures whose growth properties have been optimized under selection pressure: since the selection value of a $(\pm)$ ensemble is determined by the geometric mean of the contributions from each component (Eq. 5.114), a mutation in the nucleotide sequence leads most readily to an advantage when it increases the selection values of both (+) and (-) phenotypes by the same amount. This condition is almost always realized in molecules with high internal symmetry, since the (+) and (-) strands hardly differ from one another. The optimization law (Eq. 5.114) thus predicts the formation of highly symmetrical sequences, and this is indeed what is found in practice.

Using MDV-RNA as an initial master sequence, F.R. Kramer *et al.* (1974) carried out a modified serial transfer experiment. They superimposed upon the constraint of dilution a specific selection pressure by adding to the standard reaction mixture an RNA-binding dye (ethidium) which is known to inhibit the enzyme-catalysed replication of nucleic acids. The reaction conditions were chosen such that the dose of dye was not lethal for the replication, and also so that the elongation step - and not the initiation step - was the one retarded by the presence of the dye.

After 25 transfers (with constant incubation times of 10 minutes) an evolution product was isolated which was to a certain extent resistant to ethidium. The increase in the rate of synthesis, shown in Fig. 13.8, is clearly to be attributed to an increase in the proportion of dye-resistant RNA molecules in the course of the experiment. A quantitative evaluation of the results showed that at the end of the experiment the RNA population consisted to 99% of mutants of the original master sequence.

Since the nucleotide sequence of the master copy at the beginning of the experiment was known, the genotypic changes accompanying its evolution towards dye resistance could be followed exactly.

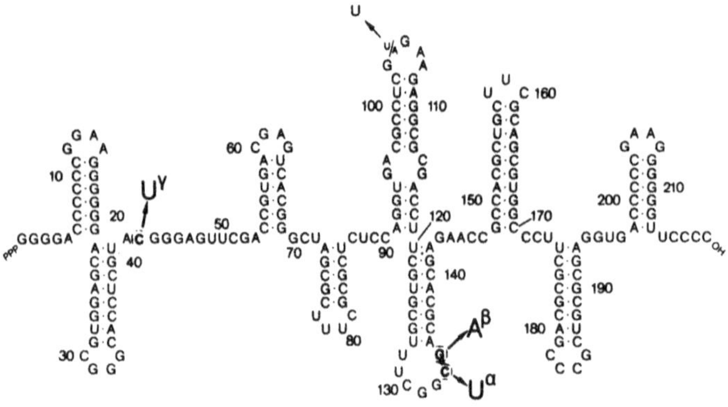

(+) Strand

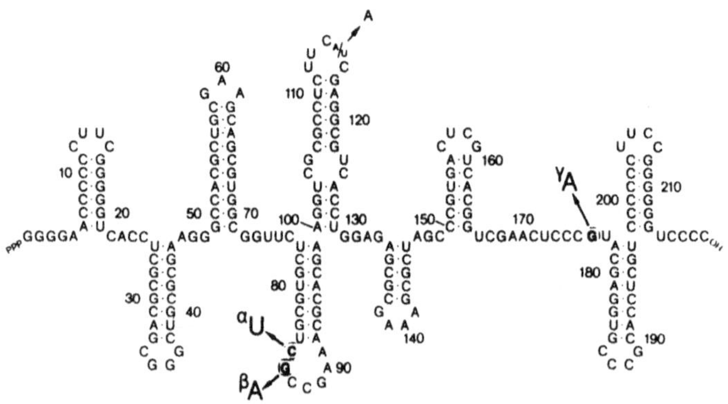

(−) Strand

Fig. 13.7. Nucleotide sequence of a typical RNA produced *de novo* by Q_β replicase under standard conditions. By convention, the (+) strand resulting from synthesis *de novo* of a (±) ensemble is the strand formed in greater concentration. The steady-state concentration ratio of a growing (±) ensemble deviates from 1:1 as soon as the two RNA components display differing replication properties (*cf.* Eq. 5.124). The better template is always present in lower concentration in the steady state.

The nucleotide sequence of the molecule shown is not quite constant. In positions 104 of the (+) strand and 115 of the (-) strand there are so-called "hot spots", *i.e.*, points where U and A are interchangeable.

S. Spiegelman and his co-workers have subjected this molecule to a serial transfer experiment in the presence of an RNA-binding dye (ethidium). The evolution product is to a certain extent resistant to the dye. The resistant RNA differs from the "wild-type" by three point mutations. The mutations did not appear simultaneously but in the order α, β, γ. The "hot spot" at position 104 of the (+) strand reverted completely to the base U, with the corresponding reversion to A at position 115 of the (-) strand. (From Kramer *et al.*, 1974.)

Repeated analysis of the evolving RNA component showed that the resistance to ethidium was brought about by three point mutations (see Fig. 13.7). It was also shown that the three mutations arose at different times, which in turn proved that the mutant RNA was not one of the mutants in the distribution about the master copy at the beginning of the experiment, but that it really did arise by mutation of the wild-type followed by natural selection.

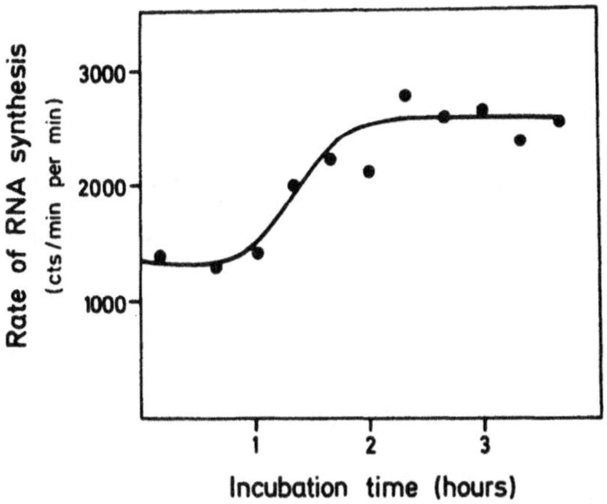

Fig. 13.8. Evolution of an RNA species resistant to ethidium. The RNA molecule in Fig. 13.7 was used by S. Spiegelman and co-workers to initiate a serial transfer experiment in the presence of ethidium bromide. The rate of RNA synthesis was determined by removing an RNA sample from each of the 24 reaction tubes and measuring its replication kinetics. The increase in the rate of RNA synthesis is due to the fact that the proportion of resistant RNA molecules per transfer in the evolving population has increased. Further details of the experiments may be found in the original paper of F.R. Kramer *et al.* (1974).

We have thus reached an important result: *Selection in the Darwinian sense can be found even at the molecular level, and can be simulated in the test-tube with isolated biological macromolecules.* This verifies the basic presuppositions of the theory which has been described in previous chapters. However, a quantitative analysis of these experiments within the framework of the theory demands the maintenance of defined and reproducible boundary conditions. We shall therefore end by presenting an experimental concept which has been designed to fulfil the conditions stated.

13.3. Experimental Perspectives: An Evolution Machine

The serial transfer experiment of Figs. 13.5 and 13.6 provides a reasonable approximation to the CP constraint discussed in Section 5.2. With each transfer the initial RNA concentration is restored, so that the total population of information carriers is, on average, kept constant during the experiment. The increase in the average excess productivity of the evolving RNA population is approximately compensated for by the steady decrease in incubation time.

The purpose of the original experiments of Spiegelman and his co-workers was to show qualitatively that selection in the Darwinian sense can be observed in the Q_β replicase system. However, the population control is by no means exact, and if the experiments are to be evaluated quantitatively then the boundary conditions must be fixed rather more precisely. We shall now take up this idea and describe a design for an "evolution machine" for nucleic acids. An analogous machine for following the evolution of bacteriophage has already been realized by Y. Husimi *et al.* (1981).

The boundary conditions are fixed exactly by replacing the discontinuous dilution of the serial transfer experiment by continuous dilution in a flow reactor.

One would have to choose whether to impose CP or CF conditions or any others. However, the experimental realization and the theoretical treatment of the CF case are much more complicated than that of the CP case; for example, to maintain CF

Fig. 13.9. Scheme of an evolution machine based upon the model system of Fig. 4.1. The reactor is operated with a flow rate of $\Phi_0(t) = dV/dt$. The solvent stream entering the reaction vessel of volume V_R contains a replicase and four kinds of energy-rich monomer with concentrations $c_y(t)$ and $c_k(t)$ respectively. These constraints are adjusted by varying the flow rates into a mixing chamber (volume V^*) at the entrance. Each component flows into the mixing chamber at a rate $\Phi_k(t)$ or $\Phi_y(t)$ from its reservoir. The solvent reservoir compensates the volume according to the relation

$$\Phi_L(t) = \Phi_0(t) - \Phi_y(t) - \sum_{k=1}^{4} \Phi_k(t)$$

The molecular species (nucleic acids) whose evolution is to be investigated are present in the reaction vessel with a concentration $x_i(t)$ at time t.

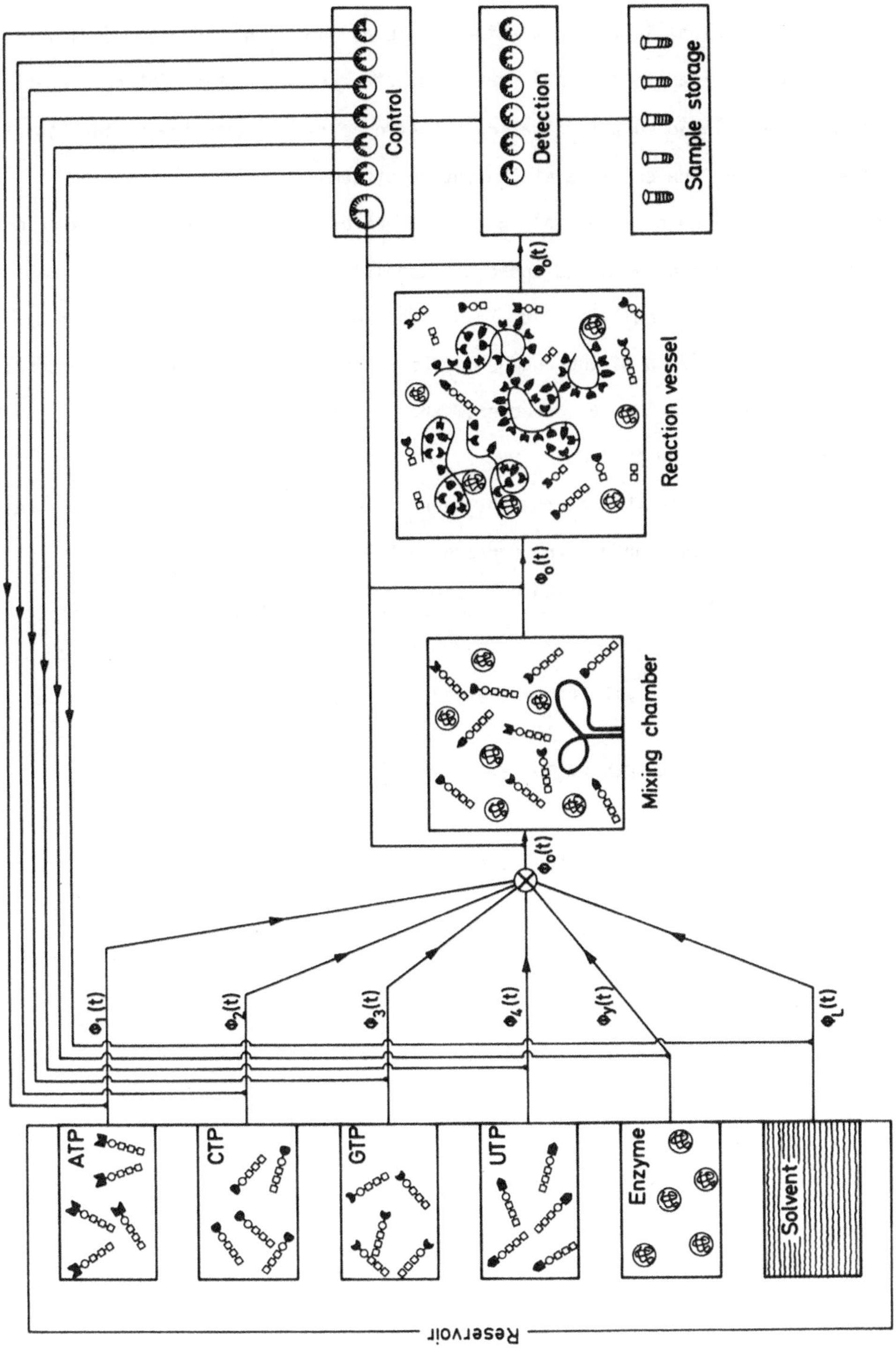
Control
Detection
Sample storage
$\Phi_o(t)$
Reaction vessel
$\Phi_o(t)$
Mixing chamber
$\Phi_o(t)$
$\Phi_1(t)$
$\Phi_2(t)$
$\Phi_3(t)$
$\Phi_4(t)$
$\Phi_y(t)$
$\Phi_L(t)$
ATP
CTP
GTP
UTP
Enzyme
Solvent
Reservoir

conditions would demand highly elaborate control mechanisms. In addition, substantial fluctuations in the population variables can occur under CF conditions, and these can in turn lead to critical states in networks of highly interdependent information units, in which a fluctuation catastrophe might cause the extinction of one component and thus the collapse of the entire system.

Our treatment which now follows is therefore restricted to the experimental realization of stationary states under CP conditions. We consider the evolution reactor shown schematically in Fig. 13.9. The reaction vessel itself consists of an open flow reactor with a mixing chamber. The reaction mixture can be continuously sampled and analysed in a detector, and samples can be collected for further characterization of the products.

Out of a mixing chamber at the entrance to the reaction vessel flows a stream of solvent which contains, apart from certain co-factors and other components of the standard reaction mixture required by the Q_β replicase system (*cf.* Fig. 13.3):

(a) energy-rich monomers with concentrations m_A, m_U, m_G and m_C,

(b) Q_β replicase with concentration y.

The control unit, coupled to the detector, regulates the component flow rates Φ_y and Φ_k and the global dilution flux Φ_0. It thus controls the concentrations of all the essential components of the reaction solution. We shall now examine more closely the process of regulation for the attainment of CP conditions.
The number z_i of information carriers I_i in a reaction volume V_R is

$$z_i(t) = x_i(t)V_R \tag{13.1}$$

while the total number of all information carriers is

$$Z(t) = \sum_i z_i(t) = \sum_i x_i(t)V_R = c(t)V_R \tag{13.2}$$

If at time t=0 the flow reactor is set in operation with a flux of

$$\Phi_0 = \frac{dV}{dt} \tag{13.3}$$

then the effective change in the quantity Z within a time interval dt is given by

$$dZ = dZ^a - dZ^d - dZ^{\Phi} \tag{13.4}$$

The individual quantities have the following meaning (where by definition $v = dZ/dt$):

(a) dZ^a is the increase in the number of information carriers in consequence of autocatalytic reactions, *i.e.*,

$$dZ^a = \sum_i dz_i^a = \sum_i v_i^a(t)dt = v^a(t)dt \tag{13.5}$$

(b) dZ^d is the decrease in this number due to degradation, *i.e.*,

$$dZ^d = \sum_i dz_i^d = \sum_i v_i^d(t)dt = v^d(t)dt \tag{13.6}$$

(c) dZ^{Φ} is the decrease in this number due to dilution, *i.e.*,

$$dZ^{\Phi} = \sum_i dz_i^{\Phi} = \sum_i x_i(t)\Phi_0(t)dt = c(t)\Phi_0(t)dt \tag{13.7}$$

The total concentration c(t) thus satisfies the differential equation

$$\frac{dc}{dt} = \frac{1}{V_R}\left(v^a(t) - v^d(t) - c(t)\Phi_0(t)\right) \tag{13.8}$$

and the solution is

$$c(t) = e^{-\frac{V(t)}{V_R}}\left(c(0) + \frac{1}{V_R}\int_0^t [v^a(\tau) - v^d(\tau)]\, e^{\frac{V(\tau)}{V_R}}\, d\tau\right) \tag{13.9}$$

where V(t), the volume of reaction mixture "exchanged" up to time t, is given by

$$V(t) = \int_0^t \Phi_0(t)dt \qquad (13.10)$$

A stationary state in which the total concentration retains its initial value c(0), *i.e.*, the CP condition, is then found if and only if

$$\frac{dc}{dt} \equiv 0 \qquad (13.11)$$

From Eq. (13.8) this state is maintained as long as the dilution flux Φ_0 can be adjusted so as to fulfil at all times the condition

$$v(t) = v^a(t) - v^d(t) \geq 0 \qquad (13.12)$$

where

$$\Phi_0(t) = \frac{1}{c(0)} v(t) \qquad (13.13)$$

We can now find out how strongly the dilution makes itself felt upon the distribution of reactants in the course of the experiment. To do this we simply "turn off" the chemical reaction in Eq. (13.9), by putting

$$v(t) = v^a(t) - v^d(t) = 0 \qquad (13.14)$$

Equation (13.9) then describes a mere dilution process:

$$c(t) = c(0)e^{-\frac{V(t)}{V_R}} \qquad (13.15)$$

This result is best illustrated by a numerical example. In the serial transfer experiment of Fig. 13.6, each transfer resulted in a dilution of the RNA population by a factor of 12.5. So at the end of the experiment, after 75 transfers, the initial concentration c(0) has been reduced $(12.5)^{75}$-fold.

Let the total "exchanged" reaction volume in the serial transfer experiment be denoted by $V_{transfer}$ and the corresponding volume in a flow experiment by V_{flow}. Their ratio is clearly expressed by the corresponding ratio for a single transfer step. Therefore, applying Eq. (13.15) to a single 12.5-fold dilution, we have

$$\frac{V(t)}{V_R} = \frac{V_{flow}}{V_{transfer}} = \ln \frac{c(0)}{c(t)} = 2.53 \qquad (13.16)$$

Thus V_{flow} exceeds $V_{transfer}$ only by a relatively small factor.

It is also interesting to compare the duration of the reactor experiment with that of the serial transfer experiment.

Suppose that the synthesis of polymer in the reactor takes place at a template concentration sufficient to saturate each enzyme molecule. The average reaction rate v is then constant (time-independent) and the dilution with respect to the initial concentration can now be calculated as a function of the duration t_D of the run. From Eq. (13.12) we have

$$t_D = \frac{V(t_D)}{V_R}\,\tau \qquad (13.17)$$

where τ is the average replicase-catalysed generation time of an information carrier. Bearing in mind that the incorporation of nucleotides into the product of evolution in the serial transfer experiment is about 2.6 times faster than into the original Q_β phage RNA, we can deduce for τ a value of approximately 3.4 min. If the overall dilution factor is

$$\frac{V(t_D)}{V_R} = (12.5)^{75} \qquad (13.18)$$

then from Eq. (13.17) we deduce a run duration of t_D = 640 min. By way of comparison, the original serial transfer experiment lasted 760 min. The difference provides a measure of the deviation between the experimental conditions of the reactor and those of the serial transfer experiment. The latter has been allowed more time than necessitated by CP conditions and has led, for this reason, to an increasing (non-stationary) rate of RNA production (*cf.* Fig. 13.6).

We have thus shown explicitly how the dilution flux Φ_0 may be used to impose certain required constraints, *e.g.* CP or CF conditions. We have also deduced that the exchange volume in a flow reactor is not much greater than the exchange volume in a typical serial transfer experiment, so that wastage of biological material (*e.g.* enzyme) in a flow reactor may be kept within acceptable limits. It can further be shown that in such an experiment a knowledge of the monomer input rates would in itself be sufficient to draw conclusions about the mechanism of nucleic acid reproduction and selection. However, the similarity between the CP flow reactor and the serial transfer experiment, in respect both of duration and of solvent consumption, suggests that the qualitative features of the flow reactor are already manifested in the much simpler serial transfer experiment.

14. Conclusions: The Logic of the Origin of Life

Let us summarize the logical pathway which we have followed in setting up our theory:

(a) The probability that a genetically integrated system will assemble spontaneously by pure chance is vanishingly small. This is true even of the chance synthesis of a small enzyme molecule optimally suited for some biological function.

(b) The first living system must therefore have been the result of a process of material self-organization, based on the selection and evolution, in the Darwinian sense, of biological macromolecules.

(c) If certain physical prerequisites are met then natural selection automatically takes place at the molecular level, leading to the generation of biological information.

(d) In a Darwinian system only a limited quantity of information can be transferred reproducibly from one generation to the next. This information threshold depends upon the accuracy of single symbol transmission.

(e) The information threshold at the inception of life could be crossed only by co-operative, information-stabilizing mechanisms. Therefore alongside the Darwinian principle there must be a further principle of self-organization of matter governing the transition from the non-living to the living.

(f) The logically simplest principle of self-organization of matter with information-stabilizing properties is the hypercycle. On account of their non-linear selection characteristics hypercycles, in constrast to Darwinian systems, evolve to the complete exclusion of their competitors.

(g) Compartmentation and individuation of hypercycles ends the non-linear nucleation phase of life and introduces the divergent phase of biological evolution.

Based on the above scheme, the flow diagram of Fig. 14.1 gives a summary of the physico-chemical steps in the transition from the non-living to the living as we see them today.

(1) PHASE OF CHEMICAL EVOLUTION: We know fairly accurately how under the conditions of the primitive earth all the chemical substances which provide the material basis of living systems could have formed. Particularly important among these are the precursors and monomers of the biological macromolecules, such as amino acids, purines, pyrimidines, phosphates, saccharides etc. This phase of chemical evolution must have been strongly divergent, since all conceivable substances allowed by the laws of chemical thermodynamics and kinetics could - and presumably did - form spontaneously.

It has been shown experimentally that the spontaneous condensation of biological macromolecules from their chemical monomers leads to random sequences.* None the less, many such random proteins were found to possess some (admittedly low) catalytic activity (*cf.* Fox, 1973). From this result it has often been surmised that the nucleation of life was inevitable as soon as a sufficiently large number of catalytic proteins was present in the prebiotic environment. However, our analysis has revealed that living systems could not have emerged merely by the random generation of proteins: a process of selective optimization must also have come into play.

(2) PHASE OF MOLECULAR SELF-ORGANIZATION: Molecular selection and evolution in the Darwinian sense presuppose a general capacity for self-replication, a property possessed by nucleic acids but not by proteins. Even though proteins seem to form more readily under prebiotic conditions than do nucleic acids, the first evolutionary

---

*This is unaffected by the slight selectivity, probably due to weak Van der Waals interactions, seen when amino acids are spontaneously incorporated into polypeptides (Harada and Fox, 1965; Steinman and Cole, 1967).

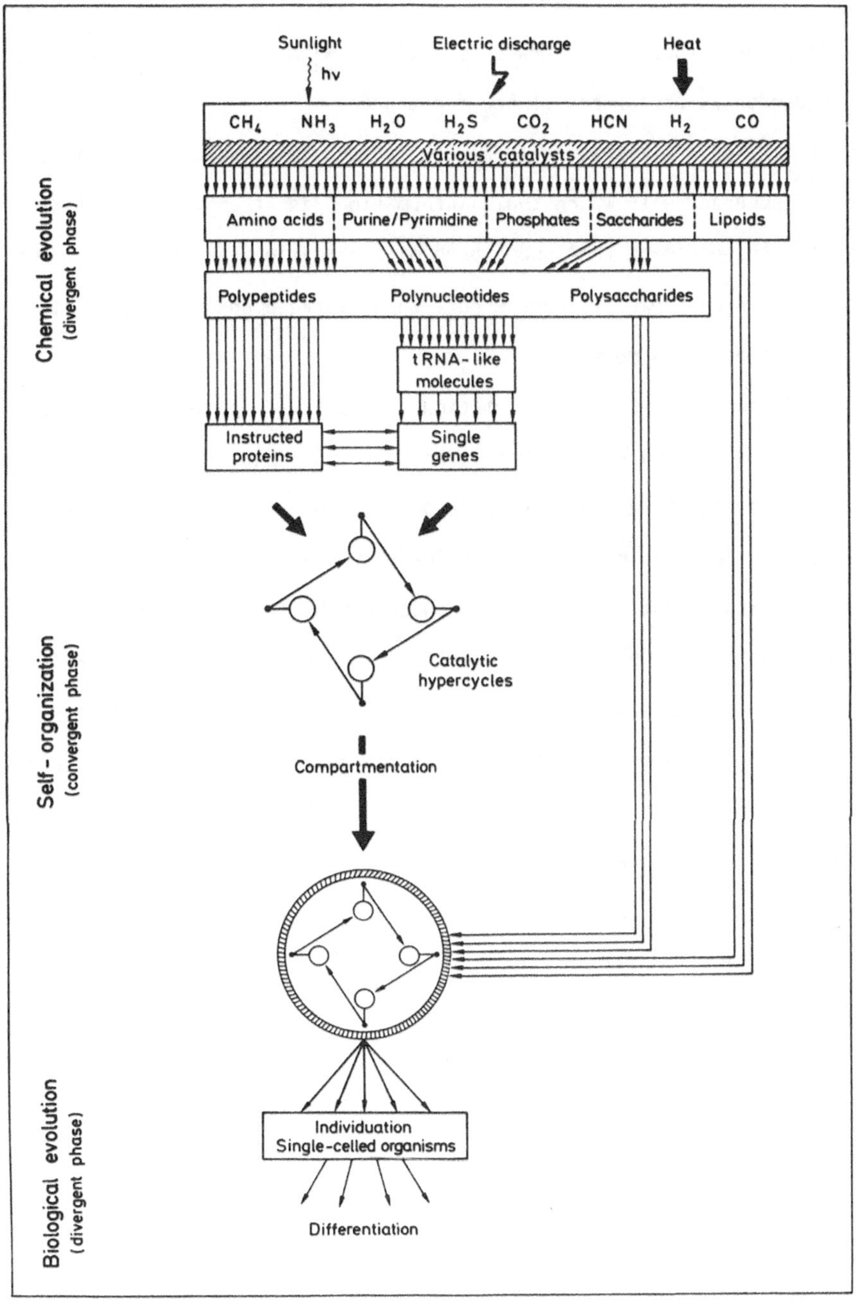

Fig. 14.1. The most important steps in the transition from the non-living to the living. (After Eigen, 1976.)

optimization of biological macromolecules may be presumed to have taken place among the nucleic acids. The ability of RNA molecules to express themselves phenotypically, and thus to interact selectively with their environment, qualifies them to be regarded as precursors of the first gene.

The evolution of nucleic acids is always subject to a limitation, placed upon it by the limited accuracy with which their information is transmitted. Experimental data lead to the estimate that under prebiotic conditions on average not more than 100 nucleotides can be copied without error, which corresponds roughly to the chain length of present-day tRNA molecules.

Further evolution of the early information carriers was possible only with the help of catalysts optimized for highly accurate copying. The primordial proteins present with random sequences could however only be optimized once they were encoded in nucleic acids. Yet the information required for the construction of the necessary translation apparatus greatly exceeds that which can be stored in a single gene with only 100 nucleotides. On the other hand, the stable co-existence of several non-related genes is impossible in a non-coupled system, where inevitably only a uniform class of molecular species and therefore only a small amount of information can survive. The co-existence of non-related information carriers could only be brought about by a new co-operative mechanism.

Such information-stabilizing co-existence is an inherent characteristic of hypercycles in general, and in particular of the catalytic hypercycle, in which proteins and nucleic acids are cyclically coupled. A hypercycle has an information capacity greater than that of a single gene and is thus equipped for the nucleation and further evolution of a complex machinery of replication and translation.

All hypercycles are optimized by natural selection. However, on account of their non-linear growth properties, hypercyclically organized systems lead inevitably to a "once-for-all-time" decision in their structure and function: as soon as a hypercycle has reached macroscopic population numbers it cannot be displaced even by selectively advantageous competitors. Even if several code schemes and/or stereospecificities are *a priori* equally likely to be found, the non-linear growth law reinforces fluctuations and will always ensure that one hypercycle becomes established and the others die out. The phase of molecular self-organization therefore must have been strongly *convergent*.

(3) PHASE OF BIOLOGICAL EVOLUTION: Since compartmentation and individuation of hypercycles are selectively advantageous, the formation of protocells was inevitable at the end of the phase of molecular self-organization. This in turn ended the non-linear selection régime of the uncompartmented hypercycles and ushered in the new, *divergent* phase of biological evolution, leading from the protocell to multicellular organisms.

We started out by asking the fundamental question:

$$\text{LIFE} = \text{CHEMISTRY} + \text{PHYSICS} \quad ? \tag{14.1}$$

Can the origin of life be explained by sole recourse to a sufficiently detailed exposition of the structure of matter? On the basis of the present theory we can now give the following answer:

A complex material system was doubtless a *necessary* prerequisite for the nucleation of life. This included a prebiotic "reaction vessel" filled with all manner of complex macromolecular structures. However, we cannot conclude from this that mere complexity of matter provides a *sufficient* basis for the nucleation of life. We are rather confronted with an information problem whose solution calls for a new physical concept of natural self-organization, involving a physical formulation of Darwin's principle of natural selection. In this way the molecular theory of evolution has brought down one of the barriers separating biology from physics and chemistry.

Mathematical Appendices

A.1. The Eigenvalue Problem

We shall first formulate the eigenvalue problem in general terms and then demonstrate its solution with a concrete example (*cf.* Eves, 1966; Eltermann, 1969).

(1) EIGENVECTORS AND EIGENVALUES. Multiplying the column vector $\mathbf{x}$ by the matrix $\mathbf{A}$ leads to a new column vector

$$\mathbf{Ax} = \begin{pmatrix} a_{11}x_1 + \cdots\cdots + a_{1n}x_n \\ a_{21}x_1 + \cdots\cdots + a_{2n}x_n \\ \vdots \qquad\qquad \vdots \\ a_{n1}x_1 + \cdots\cdots + a_{nn}x_n \end{pmatrix} \qquad (A.1.1)$$

in which each element of $\mathbf{Ax}$ is a sum of n terms. In particular, a non-zero vector $\mathbf{x}$ is called an *eigenvector* of matrix $\mathbf{A}$ if the transformation by $\mathbf{A}$ leads to the same vector $\mathbf{x}$ times a scalar λ, *i.e.*, $\mathbf{x}$ is an eigenvector of $\mathbf{A}$ if

$$\mathbf{Ax} = \lambda\mathbf{x} \qquad (A.1.2)$$

The scalar factor λ is called the *eigenvalue* of $\mathbf{A}$.

We now give a method of finding the eigenvalues and eigenvectors defined above. Eq. (A.1.2) leads to

$$(\mathbf{A} - \lambda\mathbf{I})\mathbf{x} = 0 \qquad (A.1.3)$$

where $\mathbf{I}$ is the *identity matrix*. The set of homogeneous equations (A.1.3) has non-trivial solutions if and only if

$$\det(\mathbf{A} - \lambda\mathbf{I}) = 0 \qquad (A.1.4)$$

Equation (A.1.4) is called the *characteristic equation* of the matrix $\mathbf{A}$ and gives the eigenvalues λ_i as roots of a polynomial of the nth degree. Each eigenvalue is associated with an eigenvector $\mathbf{x}^i$ according to

$$(\mathbf{A} - \lambda_i \mathbf{I})\mathbf{x}^i = 0 \qquad \text{(A.1.5)}$$

and each eigenvector consists of n elements:

$$\mathbf{x}^i = \begin{pmatrix} x_1^i \\ \vdots \\ \vdots \\ x_n^i \end{pmatrix} \qquad \text{(A.1.6)}$$

All the eigenvectors can be summarized as the columns of an *eigenvector matrix* $\mathbf{X}$

$$\mathbf{X} \equiv (\mathbf{x}^1 \ldots \mathbf{x}^i \ldots \mathbf{x}^n) = \begin{pmatrix} x_1^1 \ldots\ldots x_1^i \ldots\ldots x_1^n \\ \vdots \qquad \vdots \qquad \vdots \\ x_n^1 \ldots\ldots x_n^i \ldots\ldots x_n^n \end{pmatrix} \qquad \text{(A.1.7)}$$

where the index i denotes the ith eigenvector. With this notation Eq. (A.1.5) can be rewritten as

$$\mathbf{AX} = (\lambda_1 \mathbf{x}^1 \ldots\ldots \lambda_i \mathbf{x}^i \ldots\ldots \lambda_n \mathbf{x}^n) \qquad \text{(A.1.8)}$$

The right-hand side of this equation is equivalent to

$$\mathbf{X} \begin{pmatrix} \lambda_1 & 0 \ldots\ldots\ldots\ldots 0 \\ 0 & \lambda_2 & \vdots \\ \vdots & & 0 \\ 0 \ldots\ldots\ldots\ldots 0 & & \lambda_n \end{pmatrix} \equiv \mathbf{XL} \qquad \text{(A.1.9)}$$

in which $\mathbf{L}$ is a *diagonal matrix* made up of the eigenvalues λ_i (*N.B.* The correct order is **XL** and not **LX**!). Combining Eq. (A.1.8) and (A.1.9) gives

$$\mathbf{AX} = \mathbf{XL} \tag{A.1.10}$$

or, after multiplication by the inverse matrix $\mathbf{X}^{-1}$,

$$\mathbf{X}^{-1}\mathbf{AX} = \mathbf{X}^{-1}\mathbf{XL} = \mathbf{L} \tag{A.1.11}$$

In general, a transformation of this kind with a non-singular matrix **X** is called an *affine transformation.*

(2) EXAMPLE. Find the eigenvalues and eigenvectors of the matrix

$$\mathbf{A} = \begin{pmatrix} 2 & 1 & 0 \\ 1 & 2 & 1 \\ 0 & 1 & 2 \end{pmatrix} \tag{A.1.12}$$

The eigenvalues are solutions of the characteristic equation

$$\det(\mathbf{A} - \lambda\mathbf{I}) = 0 \tag{A.1.13}$$

Expanding the determinant leads to a polynomial of the third order

$$\det(\mathbf{A} - \lambda\mathbf{I}) = (2 - \lambda)^3 - 2(2 - \lambda) = 0 \tag{A.1.14}$$

with the solutions

$$\lambda_1 = 2 - \sqrt{2}, \qquad \lambda_2 = 2, \qquad \lambda_3 = 2 + \sqrt{2} \tag{A.1.15}$$

We now calculate the corresponding eigenvectors. To λ_1 belongs $\mathbf{x}^1$, thus

$$(\mathbf{A} - \lambda_1 \mathbf{I})\mathbf{x}^1 = 0 \qquad \text{(A.1.16)}$$

This matrix equation represents a set of homogeneous equations, from which the elements of $\mathbf{x}^1$ can be calculated:

$$\begin{aligned} \sqrt{2}\, x_1^1 + 1\, x_2^1 + 0\, x_3^1 &= 0 \\ 1\, x_1^1 + \sqrt{2}\, x_2^1 + 1\, x_3^1 &= 0 \\ 0\, x_1^1 + 1\, x_2^1 + \sqrt{2}\, x_3^1 &= 0 \end{aligned} \qquad \text{(A.1.17)}$$

Since we are looking for a non-zero vector, and since any multiple of a solution vector is also a solution, we can fix one element of $\mathbf{x}^1$ arbitrarily. Let us set $x_1^1 = 1$. With this, the remaining elements of $\mathbf{x}^1$ are determined:

$$x_1^1 = 1, \; x_2^1 = -\sqrt{2}, \; x_3^1 = 1 \qquad \text{(A.1.18)}$$

The eigenvectors $\mathbf{x}^2$ and $\mathbf{x}^3$ are found in a similar way. For the eigenvector matrix $\mathbf{X}$ we obtain finally

$$\mathbf{X} = \begin{pmatrix} 1 & 1 & 1 \\ -\sqrt{2} & 0 & \sqrt{2} \\ 1 & -1 & 1 \end{pmatrix} \qquad \text{(A.1.19)}$$

A.2. Linear Stability Analysis

Consider a set of differential equations:

$$\dot{x}_i = G_i(x_1,\dots\dots,x_N) \qquad (i = 1,\dots\dots,N) \tag{A.2.1}$$

Further, let $x_1=q_1(t),\dots\dots,x_N=q_N(t)$ be solutions of Eqs. (A.2.1). These are determined uniquely by the initial values of x_i at time $t=t_0$. The $q_i(t)$ describe a trajectory in $\mathbf{q}$-space.

We now investigate to what extent the trajectories of a "particle" in $\mathbf{q}$-space depend upon fluctuations in the initial conditions. We denote a given trajectory $\mathbf{q}(t)$ as *stable* when other trajectories $\mathbf{p}(t)$, which at time t_0 were adjacent to $\mathbf{q}(t_0)$, remain adjacent to $\mathbf{q}(t)$ at all time $t \geq t_0$.

Thus, if at time $t=t_0$ all solutions start in close proximity to $\mathbf{q}(t_0)$, the criterion of stability demands that at every subsequent point in time all

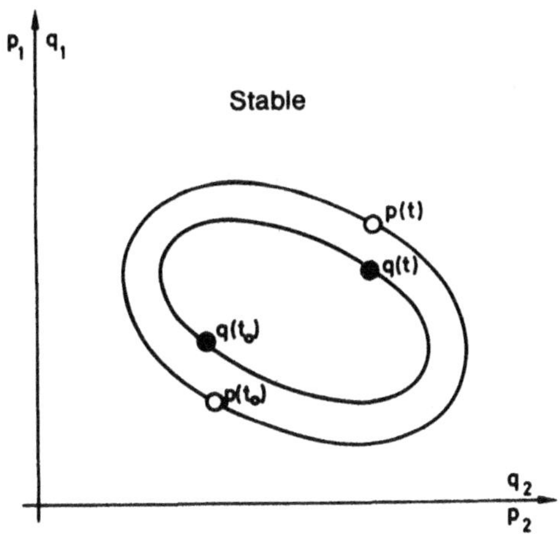

Fig. A.2.1. Example of a stable trajectory. Corresponding points on the trajectories $\mathbf{p}$ and $\mathbf{q}$ always move within each other's proximity.

trajectories remain close to one another (see Fig. A.2.1). If as $t \to \infty$ this condition is not even intermittently fulfilled, then we call $\mathbf{q}(t)$ *unstable*.

As a special case of our definition of stability we investigate the situation in which two adjacent trajectories $\mathbf{q}(t)$ and $\mathbf{p}(t)$ come arbitrarily close to each other in the course of time, so that the distance between them vanishes:

$$\lim_{t\to\infty} |\mathbf{q}(t) - \mathbf{p}(t)| = 0 \qquad \text{(A.2.2)}$$

In this case we call the trajectory $\mathbf{q}(t)$ *asymptotically stable* (see Fig. A.2.2).

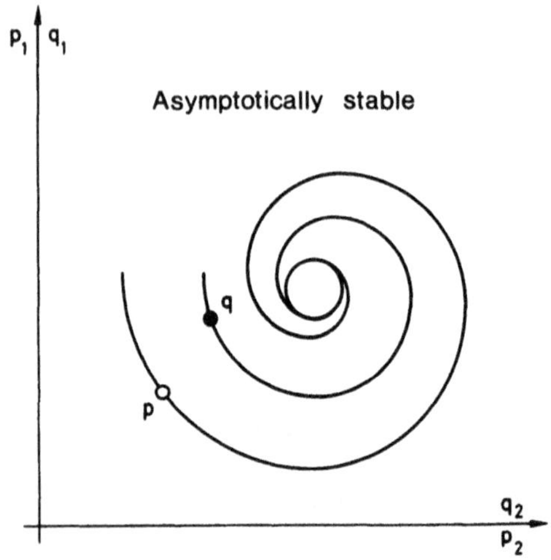

Fig. A.2.2. Example of an asymptotically stable trajectory. Corresponding points on the trajectories $\mathbf{q}$ and $\mathbf{p}$ move towards each other in the course of time so that in the long term the distance between $\mathbf{q}$ and $\mathbf{p}$ becomes vanishingly small.

(1) LOCAL STABILITY. We now give a procedure for deciding whether a given trajectory $\mathbf{q}(t)$ is asymptotically stable. We proceed from a neighbouring trajectory $\mathbf{r}(t)$

$$\mathbf{r}(t) = \mathbf{q}(t) + \boldsymbol{\xi}(t) \qquad \text{(A.2.3)}$$

which differs from $\mathbf{q}(t)$ by a small quantity $\boldsymbol{\xi}(t)$. According to our definition (Eq. A.2.2) the trajectory $\mathbf{q}(t)$ is asymptotically stable if

$$\lim_{t\to\infty} \boldsymbol{\xi}(t) = 0 \qquad \text{(A.2.4)}$$

The relation (A.2.3) can also be written in component notation

$$r_i(t) = q_i(t) + \xi_i(t) \qquad (i = 1,\ldots\ldots,N) \tag{A.2.5}$$

Inserting Eq. (A.2.5) into Eq. (A.2.1) gives

$$\dot{q}_i(t) + \dot{\xi}_i(t) = G_i(\{q_j(t)+\xi_j(t)\}) \qquad (i,j = 1,\ldots\ldots,N) \tag{A.2.6}$$

Since we have assumed that all $\xi_i(t)$ are small, we can expand the right-hand side of Eq. (A.2.6) into the well-known *Taylor series*

$$f(x_0+h) = f(x_0) + \frac{f'(x_0)}{1!}h + \frac{f''(x_0)}{2!}h^2 + \ldots\ldots \tag{A.2.7}$$

and ignore all terms after the second (hence *linear* stability analysis). From

$$G_i(\{q_j^0+\xi_j\}) = G_i(\{q_j^0\}) + \sum_k \frac{\partial G_i}{\partial \xi_k}\bigg|_0 \xi_k \tag{A.2.8}$$

follows

$$\dot{q}_i^0 + \dot{\xi}_i = G_i(\{q_j^0\}) + \sum_k \frac{\partial G_i}{\partial \xi_k}\bigg|_0 \xi_k \tag{A.2.9}$$

and from this follows

$$\dot{\xi}_i(t) = \sum_k \frac{\partial G_i}{\partial \xi_k}\bigg|_0 \xi_k(t) \tag{A.2.10}$$

As a rule it is difficult to solve the set of differential equations (A.2.10), since q_j and thus $dG/d\xi$ will usually be time-dependent. More tractable is the case in which the trajectory $\mathbf{q}$ consists solely of a point, so that all q_j are constant. For this case Eq. (A.2.5) gives

$$\frac{\partial G_i}{\partial \xi_k} = \frac{\partial G_i}{\partial x_k} \tag{A.2.11}$$

and we can set

$$\frac{\partial G_i}{\partial x_k} = a_{ik} \tag{A.2.12}$$

in which all the a_{ik} are constant. Summarizing all a_{ik} in a matrix

$$\mathbf{A} = (a_{ik}) \tag{A.2.13}$$

we now rewrite the set of differential equations (A.2.10) in the form

$$\dot{\boldsymbol{\xi}} = \mathbf{A}\boldsymbol{\xi} \tag{A.2.14}$$

This is a set of first-order linear differential equations with constant coefficients, and it can be solved by the standard methods described in Appendix A.1.

We find the eigenvalues of the matrix $\mathbf{A}$ by solving the characteristic equation

$$\det(\mathbf{A} - \lambda \mathbf{I}) = 0 \tag{A.2.15}$$

This may be expanded to give a polynomial of the Nth order:

$$f(\lambda) \equiv C_0\lambda^N + C_1\lambda^{N-1} + \ldots\ldots + C_N = 0 \tag{A.2.16}$$

If the real parts of all λ's are negative, all perturbations decay exponentially in time, so that the solution is asymptotically stable. As soon as one eigenvalue with a positive real part appears, we can always find an initial trajectory which deviates from $\mathbf{q}(t)$, so that the solution is unstable. Cases in which no λ is positive but some λ can be zero show so-called *marginal* stability. For a comprehensive and elementary introduction into the problems of stability theory see H. Haken (1977).

(2) EXAMPLE. Carry out a linear stability analysis for the selection equation (5.21):

$$\dot{x}_i = G_i(x_1,\ldots\ldots,x_N) = (W_i-\overline{E})x_i \tag{A.2.17}$$

There are N stationary states, each characterized by an N-tuplet of numbers

$$\mathbf{x}^{S_1} = \begin{pmatrix} C \\ 0 \\ \vdots \\ \vdots \\ \vdots \\ 0 \end{pmatrix},\ldots,\mathbf{x}^{S_i} = \begin{pmatrix} 0 \\ \vdots \\ 0 \\ C \\ 0 \\ \vdots \\ 0 \end{pmatrix},\ldots,\mathbf{x}^{S_N} = \begin{pmatrix} 0 \\ \vdots \\ \vdots \\ \vdots \\ 0 \\ C \end{pmatrix} \tag{A.2.18}$$

where for each state S_i the relation

$$W_i = \overline{E} \tag{A.2.19}$$

applies. Since in the stationary state the trajectories are time-independent, we can find the matrix $\mathbf{A} = (a_{ik})$ using the relation (A.2.12). From Eqs. (A.2.17) and (A.2.19) we obtain:

$$\frac{\partial G_i}{\partial x_i} = a_{ii} = W_i - \overline{E} - \frac{\partial \overline{E}}{\partial x_i} x_i = W_i - \overline{E} - \frac{1}{c} E_i x_i^s = - W_i \tag{A.2.20}$$

and

$$\frac{\partial G_k}{\partial x_k} = a_{kk} = W_k - \overline{E} - \frac{\partial \overline{E}}{\partial x_k} x_k = W_k - W_i \qquad \text{for all } k \neq i \tag{A.2.21}$$

and

$$\frac{\partial G_k}{\partial x_i} = a_{ki} = 0 \qquad \text{and} \qquad \frac{\partial G_i}{\partial x_k} = a_{ik} = - E_k \qquad \text{for } i \neq k \tag{A.2.22}$$

From the coefficients a_{ik} we can construct the matrix $\mathbf{A}$, whose eigenvalues in turn give the characteristic equation

$$(W_i+\lambda) \prod_{k \neq i} (W_k-W_i-\lambda) = 0 \tag{A.2.23}$$

which has the solutions

$$\lambda_i = - W_i \qquad \text{and} \qquad \lambda_k = W_k - W_i \qquad \text{for all } k \neq i \tag{A.2.24}$$

For a state to be stable all λ_i ($i = 1,\ldots\ldots,N$) must be negative. Equation (A.2.24) shows that this is fulfilled only by one state S_m, namely, that for which

$$W_m = \max_j W_j \tag{A.2.25}$$

All other (N-1) stationary states are unstable.

A.3. The Method of Lagrange Multipliers

We are given the function

$$f = f(x_1,\ldots\ldots,x_N) \tag{A.3.1}$$

and the task of finding its extreme values. As long as the variables x_i are all independent of one another, the extremum condition

$$\delta f = \sum_{i=1}^{N} \frac{\partial f}{\partial x_i}\,\delta x_i = 0 \tag{A.3.2}$$

holds, in which δx_i are the variations in the x_i variables. On the assumption that all x_i are independent, x_i can vary freely, *i.e.*, δx_i can adopt any value. Thus Eq. (A.3.2) holds only if each coefficient of the δx_i vanishes:

$$\frac{\partial f}{\partial x_i} = 0 \qquad \text{(for all i)} \tag{A.3.3}$$

The conditions (A.3.3) are only valid for systems free of constraint.

We now let the variables x_i be mutually dependent, in that they for example obey the CP constraint

$$\sum_{i=1}^{N} x_i = c = \text{const.} \tag{A.3.4}$$

The variables x_i can no longer change independently and the extremum principle (A.3.3) is no longer applicable. We must therefore generalize the above procedure so as to include the case in which the extreme values of the function f are to be found under restricting constraints.

To do this we first write the constraint (A.3.4) in the general form

$$g(x_1,\ldots\ldots,x_N) = 0 \qquad (A.3.5)$$

so that

$$\delta g = \sum_{i=1}^{N} \frac{\partial g}{\partial x_i} \delta x_i = 0 \qquad (A.3.6)$$

We now multiply δg by an undetermined parameter ξ and add the product to δf. From this procedure the function

$$G = \delta f + \xi \delta g \qquad (A.3.7)$$

results. Inserting the explicit expressions for δg and δf we obtain finally

$$G = \sum_i \left(\frac{\partial f}{\partial x_i} + \xi \frac{\partial g}{\partial x_i}\right) \delta x_i = 0 \qquad (A.3.8)$$

On account of the constraint (A.3.5) we have only N-1 independent variables. Without restricting the generality, let the dependent variable be x_N. By analogy with Eq. (A.3.3) we now require that

$$\frac{\partial f}{\partial x_N} + \xi \frac{\partial g}{\partial x_N} = 0 \qquad (A.3.9)$$

From this relationship we can determine the parameter ξ:

$$\xi = - \frac{\partial f}{\partial x_N} \frac{\partial x_N}{\partial g} \qquad (A.3.10)$$

The remaining N-1 variables are independent of each other and can vary freely. Considering Eq. (A.3.10) we see that Eq. (A.3.8) is satisfied only when the coefficients of the variation terms δx_i vanish:

$$\frac{\partial f}{\partial x_i} + \xi \frac{\partial g}{\partial x_i} = 0 \qquad (i = 1,\ldots\ldots,N-1) \qquad (A.3.11)$$

If we now combine Eqs. (A.3.11) with Eq. (A.3.10) we obtain for *all* N variables x_i

$$\frac{\partial f}{\partial x_i} + \xi \frac{\partial g}{\partial x_i} = 0 \qquad (i = 1,\ldots\ldots,N) \qquad (A.3.12)$$

The parameter ξ is called the *Lagrange multiplier*. The method can easily be extended to the general case of several constraints. It is merely necessary to introduce a further multiplier for each additional constraint.

A.4. Stochastic Processes

In the following treatment some of the important mathematical relationships in Chapter 6 will be deduced. For further details see A.F. Bartholomay (1958a) and J.A. Rosanov (1975).

(1) DERIVATION OF THE KOLMOGOROVIAN EQUATIONS (6.12) AND (6.13):

We start from Eq. (6.7)

$$p_{ik}(t+\tau) = \sum_j p_{ij}(\tau)p_{jk}(t) \tag{A.4.1}$$

From this follows

$$p_{ik}(t+\Delta t) = \sum_j p_{ij}(\Delta t)p_{jk}(t) = \sum_j p_{ij}(t)p_{jk}(\Delta t) \tag{A.4.2}$$

If we now insert the asymptotic expressions (6.9) and (6.10) we obtain

$$\frac{p_{ik}(t+\Delta t)-p_{ik}(t)}{\Delta t} = \sum_j \left[q_{ij}+\frac{o(\Delta t)}{\Delta t}\right] p_{jk}(t) = \sum_j p_{ij}(t)\left[q_{jk}+\frac{o(\Delta t)}{\Delta t}\right] \tag{A.4.3}$$

For $\Delta t \rightarrow 0$ the summation terms of Eq. (A.4.3) converge to

$$\sum_j q_{ij}p_{jk}(t) = \lim_{\Delta t \to 0} \sum_j \left[q_{ij}+\frac{o(\Delta t)}{\Delta t}\right] p_{jk}(t) \tag{A.4.4}$$

and

$$\sum_j q_{jk}p_{ij}(t) = \lim_{\Delta t \to 0} \sum_j \left[{}_{jk}+\frac{o(\Delta t)}{\Delta t}\right] p_{ij}(t) \tag{A.4.5}$$

The corresponding limit of the left-hand side of Eq. (A.4.3)

$$\lim_{\Delta t \to 0} \frac{p_{ik}(t+\Delta t)-p_{ik}(t)}{\Delta t} = \dot{p}_{ik}(t) \tag{A.4.6}$$

is the required derivative of $p_{ik}(t)$; equating Eqs. (A.4.4) and (A.4.5) with Eq. (A.4.6) leads to Eqs. (6.12) and (6.13) respectively.

(2) DERIVATION OF THE TRANSITION PROBABILITY EXPRESSIONS (6.20) AND (6.21):

(a) *Probability of a single birth event:* We wish to know the probability of k successes in m trials with the individual probability

$$p_f = \frac{\Delta f}{z} = F\Delta t + o(\Delta t) \tag{A.4.7}$$

We insert p_f into the *Bernoulli formula*

$$B(k,m,p) = \binom{m}{k} p^k (1-p)^{m-k} \tag{A.4.8}$$

and obtain

$$B(k,m,p_f) = \binom{m}{k}\Big[F\Delta t+o(\Delta t)\Big]^k\Big[1-(F\Delta t+o(\Delta t))\Big]^{m-k} \tag{A.4.9}$$

We now expand the right-hand side of this equation and obtain

$$B(k,m,p_f) = \binom{m}{k}\Big[(F\Delta t)^k + k(F\Delta t)^{k-1}o(\Delta t) + \frac{k(k-1)}{2}(F\Delta t)^{k-2}[o(\Delta t)]^2 + \ldots\ldots + [o(\Delta t)]^k\Big]\Big[1-g(\Delta t)\Big] \tag{A.4.10}$$

in which $g(\Delta t)$ is a polynomial in Δt with constant coefficients, so that

$$B(k,m,p_f) = \begin{cases} Fm\Delta t + o(\Delta t) & \text{for } k=1 \\ 0 \quad\;\; + o(\Delta t) & \text{for } k\geq 2 \end{cases} \tag{A.4.11}$$

i.e., the probability of the appearance of exactly one birth event (k=1) in the time interval Δt is indeed given by Eq. (6.20). The probability of more than one birth event is equal to $o(\Delta t)$, where $o(\Delta t)$ is a polynomial in Δt whose terms contain powers of Δt higher than unity.

(b) *Probability of a single death event*: Using the same procedure with the help of the relation

$$p_d = \frac{\Delta d}{z} = D\Delta t + o(\Delta t) \tag{A.4.12}$$

and the Bernoulli formula (Eq. A.4.8) we can calculate the probability that exactly one death event will take place in the time interval Δt.

(3) DERIVATION OF THE SOLUTIONS FOR THE DIFFERENTIAL EQUATIONS (6.30) AND (6.31):

We solve this set of equations by making use of a *probability generating function* Ω, which we define by

$$\Omega_i(s,t) = \sum_{k=0}^{\infty} s^k \, p_{ik}(t) \tag{A.4.13}$$

where s is an arbitrary mathematical auxiliary variable. First we differentiate the functions Ω_i partially with respect to the variables s

$$\frac{\partial \Omega_i}{\partial s} = \sum_{k=0}^{\infty} k s^{k-1} \, p_{ik}(t) \tag{A.4.14}$$

and t

$$\frac{\partial \Omega_i}{\partial t} = \sum_{k=0}^{\infty} s^k \, \dot{p}_{ik}(t) \tag{A.4.15}$$

We now insert relation (6.30) into Eq. (A.4.15) and obtain

$$\frac{\partial\Omega_i}{\partial t} = -(F+D)s \sum_{k=0}^{\infty} ks^{k-1} p_{ik}(t) + F s^2 \sum_{k=0}^{\infty} (k-1)s^{k-2} p_{i,k-1}(t) + \ldots\ldots \tag{A.4.16}$$

$$\ldots\ldots + D \sum_{k=0}^{\infty} (k+1)s^k p_{i,k+1}(t)$$

Taking into account Eq. (A.4.14) we obtain from this a partial differential equation for the generating function Ω_i, *viz.*,

$$\frac{\partial\Omega_i}{\partial t} = -(F+D)s \frac{\partial\Omega_i}{\partial s} + Fs^2 \frac{\partial\Omega_i}{\partial s} + D \frac{\partial\Omega_i}{\partial s} \tag{A.4.17}$$

or after elementary rearrangement

$$\frac{\partial\Omega_i}{\partial t} - (s-1)(Fs-D) \frac{\partial\Omega_i}{\partial s} = 0 \tag{A.4.18}$$

This is a linear, homogeneous, first-order partial differential equation with the auxiliary equation

$$\frac{dt}{ds} = -\frac{1}{(s-1)(Fs-D)} \tag{A.4.19}$$

The integration of this equation gives for the case $F \neq D$ the relation

$$C_1 = \sigma_1 e^{(F-D)t} \tag{A.4.20}$$

with

$$\sigma_1 = \frac{s-1}{Fs-D} \tag{A.4.21}$$

and for the case $F=D$ the relation

$$C_2 = \sigma_2 - Ft \tag{A.4.22}$$

with

$$\sigma_2 = \frac{1}{s-1} \tag{A.4.23}$$

where C_1 and C_2 are constants of integration.

(a) *Structure of* Ω_i *for* F≠D: A general solution of the differential equations (A.4.18) can be written as

$$\Omega_i(s,t) = \phi(C_1) = \phi\left[\sigma_1 e^{(F-D)t}\right] \tag{A.4.24}$$

For $t \rightarrow 0$ we obtain from Eqs. (A.4.13) and (6.8) the boundary condition

$$\Omega_i(s,0) = s^i \tag{A.4.25}$$

i.e.,

$$\Omega_i(s,0) = \phi(\sigma_1) = \left(\frac{\sigma_1 D-1}{\sigma_1 F-1}\right)^i \tag{A.4.26}$$

On the other hand, from Eq. (A.4.24) it follows that

$$\phi(\sigma_1 e^{(F-D)t}) = \left[\frac{\sigma_1 e^{(F-D)t}D-1}{\sigma_1 e^{(F-D)t}F-1}\right]^i = \Omega_i(s,t) \ . \tag{A.4.27}$$

so that after elementary rearrangement we obtain

$$\Omega_i(s,t) = \left[\frac{(De^{(F-D)t}-D) - (De^{(F-D)t}-F)s}{(Fe^{(F-D)t}-D) - (Fe^{(F-D)t}-F)s}\right]^i \tag{A.4.28}$$

Expansion of the right-hand side of this equation in powers of s leads to values for the coefficients of s^k, which by comparison with Eq. (A.4.13) yield ultimately the following values for the transition probabilities $p_{ik}(t)$:

$$p_{ik}(t) = \sum_{n=0}^{k \text{ or } i} (-1)^n \binom{i}{n}\binom{i+k-n-1}{k-n} F^{k-n}D^{i-n} \cdot\cdots\cdot$$
$$\cdots\cdot \left[e^{(F-D)t}-1\right]^{i+k-2n}\left[Fe^{(F-D)t}-D\right]^{-i-k+n}\left[De^{(F-D)t}-F\right]^n \tag{A.4.29}$$

where the upper summation limit is k if $0<k<i$ and i if $k\geq i$.

(b) *Structure of* Ω_i *for* F=D: A general solution of the differential equations (A.4.18) can be written as

$$\Omega_i(s,t) = \phi(C_2) = \phi(\sigma_2-Ft) \tag{A.4.30}$$

For $t\rightarrow 0$ we obtain from Eqs. (A.4.17) and (6.8) the boundary condition

$$\Omega_i(s,0) = s^i \tag{A.4.31}$$

i.e.,

$$\Omega_i(s,0) = \phi(\sigma_2) = \left(\frac{\sigma_2+1}{\sigma_2}\right)^i \tag{A.4.32}$$

On the other hand, from Eq. (A.4.30) it follows that

$$\phi(\sigma_2-Ft) = \left(\frac{\sigma_2-Ft+1}{\sigma_2-Ft}\right)^i \tag{A.4.33}$$

so that after elementary rearrangement we obtain

$$\Omega_i(s,t) = \left[\frac{Ft - (Ft-1)s}{(1+Ft) - Ft\ s}\right]^i \tag{A.4.34}$$

Expansion of the right-hand side of this equation in powers of s leads to values for the coefficients of s^k, which by comparison with Eq. (A.4.13) ultimately yield the following values from the transition probabilities $p_{ik}(t)$

$$p_{ik}(t) = \sum_{n=0}^{k \text{ or } i} (-1)^n \binom{i}{n}\binom{i+k-n-1}{k-n}(Ft)^{i+k-2n}(Ft-1)^n(Ft+1)^{-i-k+n} \tag{A.4.35}$$

where the upper summation limit is k if $0<k<i$ and i if $k\geq i$.

Bibliography

An asterisk (*) indicates that the reference has not been cited in the text, but deals closely with the problems treated in this book and is recommended for further reading. No attempt has been made at completeness.

AYALA, F. (1978): The Mechanisms of Evolution. Scientific American 239(9), 48.

BARRELL, B.G., CLARK, B.F.C. (1974): Handbook of Nucleic Acid Sequences. Joynson-Bruvvers, Oxford.

BARTHOLOMAY, A.F. (1958a): On the Linear Birth and Death Processes of Biology as Markoff Chains. Bull. Math. Biol. 20, 97.

BARTHOLOMAY, A.F. (1958b): Stochastic Models for Chemical Reactions: I. Theory of the Unimolecular Reaction Process. Bull. Math. Biol. 20, 175.

BATSCHELET, E., DOMINGO, E., WEISSMANN, C. (1976): The Proportion of Revertant and Mutant Phage in a Growing Population, as a Function of Mutant and Growth Rate. Gene 1, 27.

BELLMAN, R. (1970): Introduction to Matrix Analysis. McGraw-Hill, New York.

BIEBRICHER, C.K., EIGEN, M., LUCE, R. (1981a): Kinetic Analysis of Template-Instructed and *de novo* RNA Synthesis by Q_β Replicase. J. Mol. Biol. 148, 391.

BIEBRICHER, C.K., EIGEN, M., LUCE, R. (1981b): Product Analysis of RNA Generated *de novo* by Q_β Replicase. J. Mol. Biol. 148, 369.

BONITZ, S.G., BERLANI, R., CORUZZI, G., LI, M., MACINO, G., NOBREGA, F.G., NOBREGA, M.P., THALENFELD, B.E., TZAGOLOFF, A. (1980): Codon Recognition Rules in Yeast Mitochondria. Biochemistry 77, 3167.

BRIDSON, P.K., ORGEL, L.E. (1980): Catalysis of Accurate Poly(C)-directed Synthesis of 3′-5′-linked Oligoguanylates by Zn^{2+}. J.Mol.Biol. 144, 567.

CRICK, F.H.C. (1966): Codon-Anticodon Pairing: The Wobble Hypothesis. J. Mol. Biol. 19, 548.

CRICK, F.H.C. (1968): The Origin of the Genetic Code. J. Mol. Biol. 38, 367.

CRICK, F.H.C., BRENNER, S., KLUG, A., PIECZENIK, G. (1976): A Speculation on the Origin of Protein Synthesis. Origins of Life 7, 389.

CROMIN, J.R., MOORE, C.B. (1971): Amino Analyses of the Murchison, Murray, and Allende Carbonaceous Chondrites. Science 172, 1327.

CROW, J.F., KIMURA, M. (1970): An Introduction to Population Genetics Theory. Harper & Row, New York.

DARWIN, C. (1859): On the Origin of Species by Means of Natural Selection. Murray, London. Reprinted by Harvard University Press.

DAVIES, B.K. (1978): Rate of Polymer Formation and Entropy Production During Competitive Replication. J. Mol. Evol. 10, 325.

DOMINGO, E., FLAVELL, R.A., WEISSMANN, C. (1976): *In vitro* Site-directed Mutagenesis: Generation and Properties of an Infectious Extracistronic Mutant of Bacteriophage Q_β. Gene 1, 3.

DOMINGO, E., SABO, D., TANIGUCHI, T., WEISSMANN, C. (1978): Nucleotide Sequence Heterogeneity of an RNA Phage Population. Cell 13, 735.

DOOB, J.L. (1953): Stochastic Processes. Wiley, New York.

*EBELING, W., FEISTEL, R. (1977): Stochastic Theory of Molecular Replication Processes with Selection Character. Ann. Physik. 34, 81.

*EBELING, W., FEISTEL, R. (1979): On the Evolution of Biological Macromolecules: I. Physico-Chemical Self-Organization. Studia Biophysica 75, 131.

*EBELING, W., MAHNKE, R. (1979): Kinetics of Molecular Replication and Selection. In: Problems of Contemporary Biophysics 4, 119.

EHRENFEST, P., EHRENFEST, T. (1907): Über zwei bekannte Einwände gegen das Boltzmannsche H-Theorem. Phys. Z. 8, 311.

EIGEN, M. (1971): Self-Organization of Matter and the Evolution of Biological Macromolecules. Naturwissenschaften 58, 465.

*EIGEN, M. (1973): The Origin of Biological Information. In: The Physicist's Conception of Nature, ed. J. Mehra. Reidel, Dordrecht.

EIGEN, M., WINKLER-OSWATITSCH, R. (1975): Das Spiel. Piper-Verlag, München. English edition: Laws of the Game. Knopf, New York,1982.

EIGEN, M. (1976): Wie entsteht Information? Prinzipien der Selbstorganisation in der Biologie. Ber. Bunsenges. physik. Chemie 80, 1060. Reprinted in English as: How Does Information Originate? Principles of Biological Self-Organization. In: Advances in Chemical Physics 38, ed. S.A. Rice. Wiley, New York, 1980.

EIGEN, M., SCHUSTER, P. (1979): The Hypercycle - A Principle of Natural Self-Organization. Springer-Verlag, Heidelberg.

EIGEN, M. (1980): Zeugen der Genesis - Versuch einer Rekonstruktion der Urformen des Lebens aus ihren in den Biomolekülen hinterlassenen Spuren. In: Jahrbuch der Max-Planck-Gesellschaft 1979, ed. Generalverwaltung der MPG, Vandenhoek & Ruprecht, Göttingen.

EIGEN, M., GARDINER, W.C., SCHUSTER, P. (1980): Hypercycles and Compartments: Compartments Assist - but Do Not Replace - Hypercyclic Organization of Early Genetic Information. J. Theor. Biol. 85, 407.

*EIGEN, M., SCHUSTER, P., SIGMUND, K., WOLFF, R. (1980): Elementary Step Dynamics of Catalytic Hypercycles. Biosystems 13, 1.

*EIGEN, M., GARDINER, W.C., SCHUSTER, P., WINKLKER-OSWATITSCH, R. (1981): The Origin of Genetic Information. Scientific American, 244(4), 88.

EIGEN, M., WINKLER-OSWATITSCH, R. (1981a): Transfer-RNA: The Early Adaptor. Naturwissenschaften 68, 217.

EIGEN, M., WINKLER-OSWATITSCH, R. (1981b): Transfer-RNA, an Early Gene? Naturwissenschaften 68, 282.

ELTERMANN, H. (1969): Grundlagen der praktischen Matrizenrechnung. Bibliographisches Institut, Mannheim.

ENGEL, J., SCHWARZ, G. (1970): Kooperative Konformationsumwandlungen von linearen Biopolymeren. Angew. Chemie 12, 468; Int. Ed. 9, 389.

EPSTEIN, I.R., EIGEN, M. (1979): Selection and Self-Organization of Self-Reproducing Macromolecules under the Constraint of Constant Flux. Biophys. Chem. 10, 153.

*EPSTEIN, I.R. (1979): Coexistence, Competition and Hypercyclic Interaction in some Systems of Biological Interest. Biophysical Chem. 9, 245.

*EPSTEIN, I.R. (1979): Competitive Coexistence of Self-Reproducing Macromolecules. J. Theor. Biol. 78, 271.

EVES, H. (1966): Elementary Matrix Theory. Dover, New York.

FIERS, W., CONTRERAS, R., DUERINCK, F., HAEGEMAN, G., ISERENTANT, D., MERREGAERT, J., MIN JOU, W., MOLEMANS, F., RAEYMAEKERS, A., VAN DEN BERGHE, G., VOLEKAERT, G., YSEBAERT, M. (1976): Complete Nucleotide Sequence of Bacteriophage MS2 RNA: Primary and Secondary Structure of the Replicase Gene. Nature 260, 500.

*FEISTEL, R., EBELING, W. (1978): On the Eigen-Schuster Concept of Quasispecies in the Theory of Natural Self-Organization. Studia Biophysica 71, 139.

FISHER, R.A. (1930): The Genetical Theory of Natural Selection. Clarendon Press, Oxford.

FITCH, W.M., MARGOLIASH, E. (1967): Construction of Phylogenetic Trees. Science 155, 279.

FOX, S.W. (1965): Simulated Natural Experiments in Spontaneous Organization of Morphological Units from Protenoid. In: The Origins of Prebiological Systems, ed. S.W. Fox. Academic Press, New York.

FOX, S.W. (1973): Origin of the Cell: Experiments and Premisses. Naturwissenschaften 60, 359.

FULLER, W., HODGSON, A. (1967): Conformation of the Anticodon Loop in tRNA. Nature 215, 817.

GLANSDORFF, P., PRIGOGINE, I. (1971): Thermodynamic Theory of Structure, Stability and Fluctuations. Wiley-Interscience, New York.

*HAAR, D. ter, GREEN, C.D. (1953): The Statistical Aspect of Boltzmann's H-Theorem. Proc. Phys. Soc. London A 66, 153.

HAKEN, H. (1977): Synergetics. Springer-Verlag, Heidelberg.

HALL, B.D. (1979): Mitochondria Spring Surprises. Nature 282, 129.

HARADA, K., FOX, S.W. (1965): Thermal Polycondensation of Free Amino Acids with Polyphosphoric Acid. In: The Origins of Prebiological Systems, ed. S.W. Fox. Academic Press, New York.

HARUNA, I., NOZU, K., OHTAKA, Y., SPIEGELMAN, S. (1963): An RNA "Replicase" Induced by and Selective for a Viral RNA: Isolation and Properties. Proc. Natl. Acad. Sci. USA 50, 905.

HARUNA, I., SPIEGELMAN, S. (1965): Specific Template Requirements of RNA Replicases. Proc. Natl. Acad. Sci. USA 54, 579.

HENDRY, L.B., BRANSOME, E.D., HUTSON, M.S., CAMPBELL, L.K. (1981): First Approximation of a Stereochemical Rationale for the Genetic Code Based on the Topography and Physicochemical Properties of "Cavities" Constructed from Models of DNA. Proc. Natl. Acad. Sci. USA 78, 7440.

HIRSCH, M.W., SMALE, S. (1974): Differential Equations, Dynamical Systems and Linear Algebra. Academic Press, New York.

*HOFBAUER, J., SCHUSTER, P., SIGMUND, K., WOLFF, R. (1980): Dynamical Systems under Constant Organization II: Homogeneous Growth Functions of Degree p=2. SIAM J.Appl.Math.C. 38, 282.

*HOFBAUER, J., SCHUSTER, P., SIGMUND, K. (1981): Competition and Cooperation in Catalytic Selfreplication. J.Math.Biol. 11, 155.

HUSIMI, Y. NISHIGAKI, K., KINOSHITA, Y., TANAKA, T. (1981): Cellstat - A continuous culture system of a bacteriophage for the study of the mutation rate and the selection process at the DNA level. Rev. Sci. Instrum. 53, 517.

*INAGAKI, H. (1982): Selection under Random Mutations in Stochastic Eigen Model. Bull. Math. Biol. 44, 17.

*ISHIDA, K. (1981): Non-equilibrium Thermodynamics of the Selection of Biological Macromolecules. J. Theor. Biol. 88, 257.

ISING, E. (1925): Beitrag zur Theorie des Ferromagnetismus. Z. Physik 31, 253.

JONES, B.L., ENNS, R.H., RANGNEKAR, S.S. (1976a): On the Theory of Selection of Coupled Macromolecular Systems. Bull. Math. Biol. 38, 15.

JONES, B.L. (1976b): A Model for Macromolecular Selection in Complementary Instructing Systems. J. Math. Biol. 4, 29.

*JONES, B.L. (1977): A Solvable Self-reproductive Hypercycle Model for the Selection of Biological Macromolecules. J. Math. Biol. 4, 187.

JONES, B.L. (1977): Analysis of Eigen's Equations for Selection of Biological Molecules with Fluctuating Mutation Rates. Bull. Math. Biol. 39, 311.

*JONES, B.L. (1979): Some Selection Models in Catalytically Coupled Self-reproducing Macromolecular Systems. Bull. Math. Biol. 41, 119.

*JONES, B.L. (1979): Selection in Systems of Self-reproducing Macromolecules under the Constraint of Controlled Energy Fluxes. Bull. Math. Biol. 41, 761.

*JONES, B.L. (1979): Stability and Selection in Self-Reproducing Macromolecular Systems. Bull. Math. Biol. 41, 449.

*JONES, B.L. (1979): Some Models for Selection of Biological Macromolecules with Time-varying Constraints. Bull. Math. Biol. 41, 849.

*JONES, B.L. (1980): On the Evolution of Prebiotic Enzyme-Coupled Macromolecular Systems. Bull. Math. Biol. 42, 539.

JONES, B.L., LEUNG, H.K. (1981): Stochastic Analysis of a Nonlinear Model for Selection of Biological Macromolecules. Bull. Math. Biol. 43, 665.

*KAC, M. (1947): Random Walk and the Theory of Brownian Motion. Amer. Math. Monthly 54, 369.

KENDREW, J.C. (1961): The Three-Dimensional Structure of a Protein Molecule. Scientific American 205(12), 96.

KILPATRICK, M.W., WALKER, R.T. (1981): The Nucleotide Sequence of the $tRNA_M^{Met}$ from the Archaebacterium *Thermoplasma Acidophilum*. Nucleic Acids Res. 9, 4387.

*KLEIN, M.J. (1956): Generalization of the Ehrenfest Urn Model. Phys. Rev. 103, 17.

KLUG, A., BUTLER, P.J.G. (1978): The Assembly of a Virus. Scientific American 239(11), 52.

KOHLRAUSCH, K.W.F., SCHRÖDINGER, E. (1926): Das Ehrenfestsche Modell der H-Kurve. Phys. Z. 27, 306.

KRAMER, F.R., MILLS, D.R., COLE, P.E., NISHIHARA, T., SPIEGELMAN, S. (1974): Evolution *in vitro*: Sequence and Phenotype of a Mutant RNA Resistant to Ethidium Bromide. J. Mol. Biol. 89, 719.

KRAMERS, H.A., WANNIER, G.H. (1941): Statistics of the Two-Dimensional Ferromagnet. Part I. Physic. Rev. 60, 252.

KUHN, H. (1972): Selbstorganisation molekularer Systeme und die Evolution des genetischen Apparats. Angew. Chem. 84, 838; Int. Ed. 11, 798.

KUHN, H. (1976): Model Consideration for the Origin of Life. Naturwissenschaften 63, 68.

KUHN, H. (1977): Morphogenesis. Persistence of Organized Structures and Breakthrough of New Structures. In: Synergetics, ed. H. Haken. Springer-Verlag, Berlin.

KUHN, H., WASER, J. (1981): Molekulare Selbstorganisation und Ursprung des Lebens. Angew. Chem. 93, 495; Int. Ed. 20, 500.

*KÜPPERS, B.-O. (1975): The General Principles of Selection and Evolution at the Molecular Level. Progr. Biophys. Molec. Biol. 30, 1.

KÜPPERS, B.-O. (1979a): Some Remarks on the Dynamics of Molecular Self-Organization. Bull. Math. Biol. 41, 803.

KÜPPERS, B.-O. (1979b): Towards an Experimental Analysis of Molecular Self-Organization and Precellular Darwinian Evolution. Naturwissenschaften 66, 228.

KVENVOLDEN, K., LAWLESS, J., PERING, K., PETERSON, E., FLORES, J., PONNAMPERUMA, C., KAPLAN, I.R., MOORE, C. (1970): Evidence for Extraterrestrial Amino-acids und Hydrocarbons in the Murchison Meteorite. Nature 228, 923.

KYOGOKU, Y., LORD, R.C., RICH, A. (1969): An Infrared Study of the Hydrogen Bonding Specificity of Hypoxanthine and other Nucleic Acid Derivatives. B.B.A. 179, 10.

LABUDA, D., PÖRSCHKE, D. (1980): Multistep Mechanism of Codon Recognition by Transfer Ribonucleid Acid. Biochemistry 19, 3799.

LANDAUER, R. (1975): Inadequacy of Entropy and Entropy Derivatives in Characterizing the Steady State. Phys. Rev. A, 12, 636.

LOHRMANN, R., BRIDSON, P.K., ORGEL, L.E. (1980): Efficient Metal-Ion Catalyzed Template-Directed Oligonucleotide Synthesis. Science 208, 1464.

McQUARRIE, D.A. (1967): Stochastic Approach to Chemical Kinetics. Methuen, London.

MILLER, S.L., ORGEL, L.E. (1974): The Origins of Life on Earth. Prentice-Hall, New Jersey.

MILLS, D.R., PETERSON, R.L., SPIEGELMAN, S. (1967): An Extracellular Darwinian Experiment with a Self-Duplicating Nucleic Acid Molecule. Proc. Natl. Acad. Sci. USA 58, 217.

MILLS, D.R., KRAMER, F.R., SPIEGELMAN, S. (1973): Complete Nucleotide Sequence of a Replicating RNA Molecule. Science 180, 916.

MONOD, J. (1970): Le hasard et la nécessité. Editions du Seuil, Paris. English edition: Chance and Necessity. Knopf, New York 1972.

MOORE, W.J. (1972): Physical Chemistry. 5th Ed. Longman, New Jersey.

NICOLIS, G, PRIGOGINE, I. (1977): Self-Organization in Nonequilibrium Systems. From Dissipative Structures to Order through Fluctuations. Wiley, New York.

ONSAGER, L. (1931a): Reciprocal Relations in Irreversible Processes. I. Physical Review 37, 405.

ONSAGER, L. (1931b): Reciprocal Relations in Irreversible Processes. II. Physical Review 38, 2265.

OPARIN, A.I. (1924): Proiskhozdenie zhizny, Izd. Moskovskii Rabochii, Moscow. Reprinted in English in J.D. Bernal "The Origin of Life". Weidenfeld and Nicolson, London, 1967.

OPARIN, A.I. (1965): The Pathways of the Primary Development of Metabolism and Artificial Modeling of this Development in Coacervate Drops. The Origins of Prebiological Systems, ed. S.W. Fox. Academic Press, New York.

*ORGEL, L.E. (1968): Evolution of the Genetic Apparatus. J. Mol. Biol. 38, 381.

*ORGEL, L.E. (1979): Selection *in vitro*. Proc. R. Soc. Lond. B 205, 435.

ORÓ, J., GIBERT, J., LICHTENSTEIN, H., WIKSTROM, S., FLORY, D.A. (1971): Amino-acids, Aliphatic and Aromatic Hydrocarbons in the Murchison Meteorite. Nature 230, 105.

PÖRSCHKE, D. (1971): Cooperative Nonenzymic Base Recognition. II. Thermodynamics of the Helix-Coil Transition of Oligoadenylic + Oligouridylic Acids. Biopolymers 10, 1989.

PÖRSCHKE, D., UHLENBECK, O.C., MARTIN, F.H. (1973): Thermodynamics and Kinetics of the Helix-Coil Transition of Oligomers Containing GC Base Pairs. Biopolymers 12, 1313.

PÖRSCHKE, D. (1977): Elementary Steps of Base Recognition and Helix-Coil Transitions in Nucleic Acids. In: Molecular Biology, Biochemistry and Biophysics, Vol. 24, ed. I. Pecht and R. Rigler. Springer-Verlag, Berlin.

PRIGOGINE, I. (1945): Moderation et transformations irréversibles des systèmes ouverts. Acad. Roy. Belg., Bull. Classe. Sci. 31, 600.

RASZKA, M., MANDEL, M. (1972): Is There a Physical Chemical Basis for the Present Genetic Code? J. Molec. Evolution 2, 38.

RICH, A., KIM, S.H. (1978): The Three-dimensional Structure of Transfer RNA. Scientific American 238(1), 52.

RICHTER, P.H. (1975): Stochastic Theory of a Selection Game. Bull. Math. Biol. 37, 193.

ROSANOV, J.A. (1975): Stochastische Prozesse. Akademie-Verlag, Berlin.

SAFFHILL, R., SCHNEIDER-BERNLOEHR, H., ORGEL, L.E., SPIEGELMAN, S. (1970): *In vitro* Selection of Bacteriophage Q_β Ribonucleic Acid Variants Resistent to Ethidium Bromide. J.Mol.Biol. 51, 531.

SAXINGER, C., PONNAMERUMA, C. (1971): Experimental Investigation on the Origin of the Genetic Code. J. Molec. Evolution 1, 63.

*SCHNEIDER, F.W. (1976): Stability of Steady States in Nucleic Acid Poly d(A-T). Synthesis and the Stirred Flow Reactor. Biopolymers 15, 1.

SCHUSTER, P. (1972): Vom Makromolekül zur primitiven Zelle - die Entstehung biologischer Funktion. Chemie in unserer Zeit 6, 1. Extended version of this article (together with K. SIGMUND): From Biological Macromolecules to Protocells - The Principle of Early Evolution. In: Biophysics, eds. W. Hoppe, W. Lohmann, H. Markl and H. Ziegler. Springer-Verlag, Heidelberg 1982.

SCHUSTER, P., SIGMUND, K., WOLFF, R. (1977): Dynamical Systems under Constant Organization I. Topological Analysis of a Family of Non-Linear Differential Equations - A Model for Catalytic Hypercycles. Bull. Math. Biol. 40, 743.

*SCHUSTER, P., SIGMUND, K., WOLFF, R. (1979): Dynamical Systems under Constant Organization III: Cooperative and Competitive Behavior of Hypercycles. Journal of Differential Equs. 32, 357.

*SCHUSTER, P., SIGMUND, K., WOLFF, R. (1979): On ω-Limits for Competition Between Three Species. SIAM J.Appl. Math. 37, 49.

*SCHUSTER, P., SIGMUND, K. (1980): Self-organization of Biological Macromolecules and Evolutionary Stable Strategies. In: Dynamics of Synergetic Systems, ed. H. Haken. Springer-Verlag, Heidelberg.

*SCHUSTER, P., SIGMUND, K. (1980):A Mathematical Model of the Hypercycle. In: Dynamics of Synergetics, ed. H.Haken. Springer-Verlag, Heidelberg.

*SCHUSTER, P. (1981): Prebiotic Evolution. In: Biochemical Evolution, ed. H. Gutfreund. Cambridge University Press, Cambridge (U.K.).

SHEPHERD, J.C.W. (1981a): Periodic Correlations in DNA Sequences and Evidence Suggesting Their Evolutionary Origin in a Comma-less Genetic Code. J.Mol.Evol. 17, 94.

SHEPHERD, J.C.W. (1981b): Method to Determine the Reading Frame of a Protein from the Purine/Pyrimidine Genome Sequence and its Possible Evolutionary Justification. Proc. Natl. Acad. Sci. USA 78, 1596.

STEINMAN, G., COLE, M.N. (1967): Synthesis of Biologically Pertinent Peptides under Possible Primordial Conditions. Proc. Natl. Acad. Sci. USA 58, 735.

SUMPER, M., LUCE, R. (1975): Evidence for *de novo* Production of Self-replicating and Environmentally Adapted RNA Structures by Bacteriophage Q_β Replicase. Proc. Natl. Acad. Sci. USA 72, 162.

THOMPSON, C.J., McBRIDE, J.L. (1974): On Eigen's Theory of the Self-Organization of Matter and the Evolution of Biological Macromolecules. Math. Biosci. 21, 127.

TINOCO, I., BORER, P.N., DENGLER, B., LEVINE, M.D., UHLENBECK, O.C., CROTHERS, D.M., GRALLA, J. (1973): Improved Estimation of Secondary Structure in Ribonucleic Acids. Nature New Biology 246, 40.

TYSON, J. (1974): Competition, Selection and Evolution in Chemical Networks. In: Some Mathematical Questions in Biology, ed. S.A. Levin. AMS Press, Rhode Island.

WATSON, J.D. (1976): Molecular Biology of the Gene. 3rd Ed. Benjamin, New York.

WOESE, C.R. (1967): The Origins of the Genetic Code. Harper & Row, New York.

WOESE, C.R. (1970): Molecular Mechanics of Translation: a Reciprocating Ratchet Mechanism. Nature 226, 817.

WOESE, C.R. (1973): Evolution of the Genetic Code. Naturwissenschaften 60, 447.

WOESE, C.R. (1981): Archaebacteria. Scientific American 244(6), 94.

ZIMM, B.H., BRAGG, J.K. (1958): Theory of the One-Dimensional Phase Transition in Polypeptide Chains. J. Chem. Physics 28, 1246.

ZIMM, B.H., BRAGG, J.K. (1959): Theory of the Phase Transition between Helix and Random Coil in Polypeptide Chains. J. Chem. Physics 31, 526.

Index

H. Haken
Synergetik

Eine Einführung

Nichtgleichgewichts-Phasenübergänge und Selbstorganisation in Physik, Chemie und Biologie
Übersetzt aus dem Englischen von A. Wunderlin
2. Auflage. 1983. 160 Abbildungen. XIV, 386 Seiten
ISBN 3-540-12597-3

Dieses zunächst in Englisch, Russisch, Japanisch und in Chinesisch erschienene Buch liegt hiermit in 2. Auflage in deutscher Sprache vor. Es befaßt sich mit den kürzlich entdeckten tiefgreifenden und verblüffenden Analogien bei Selbstorganisationsvorgängen scheinbar ganz verschiedener Systeme in der Physik, Chemie, Biologie, Soziologie und anderen Gebieten.
Die Kooperation vieler Einzelteile, wie Atome, Moleküle, Zellen, Tiere oder selbst Menschen, kann räumliche, zeitliche oder funktionelle Strukturen hervorbringen. Ihre spontane Bildung aus dem Chaos heraus erinnert oft stark an Phasenübergänge.
Dieses vom Begründer der Synergetik geschriebene Buch vermittelt eine elementare Einführung in die grundlegenden Ideen und mathematischen Hilfsmittel. Zahlreiche Übungsaufgaben, Figuren und einfache Beispiele helfen zum Verständnis. Die grundlegenden Analogien werden an realistischen Beispielen demonstriert, wie Flüssigkeitsdynamik, Laser, chemische und biologische Systeme, Ökologie, Ökonomie, Soziologie und Theorien der Evolution und Morphogenese.
In der vorliegenden 2. Auflage ist ein Abschnitt über diskrete Abbildungen, die ein wichtiges Werkzeug zum Verständnis von Periodenverdoppelungen, Intermittenz und Chaos sind, hinzugekommen.

Springer-Verlag
Berlin
Heidelberg
New York
Tokyo

P. K. Kuroda

The Origin of the Chemical Elements

and the Oklo Phenomenon

1982. 24 figures, 48 tables. XI, 165 pages
ISBN 3-540-11679-6

Major developments in the field of nuclear geochemistry and cosmochemistry are reviewed in this monograph. Following a brief introduction, an historical account of the early ideas concerning the cosmic abundance of the elements and the searches made for the 'missing' elements 43 (Tc) and 61 (Pm) in nature are given. The sequence of events which culminated in the discovery of the Oklo Phenomenon (Pre-Fermi reactor), and the topics related to the synthesis of the elements in stars are then discussed as are the ideas concerning the extinct radioactivities and the discoveries of the extinct nuclides ^{129}I and ^{244}Pu.

In the final chapter on isotopic anomalies in the early solar system, the author presents an unbiased review of an area that – although dating back to the days of ancient Greek philosophers and regarded by many as the most fundamental in the entire compass of our modern science – is far from settled and is perhaps not quite ready for incorporation into textbooks.

Contents: Introduction. – Abundance of the Elements. – Elements 43 and 61 in Nature. – The Oklo Phenomenon. – Synthesis of the Elements in Stars. – Plutonium-244 in the Early Solar System. – Isotopic Anomalies in the Early Solar System. – Appendices: Appendix I: Goldschmidt's Table of the Abundance of the Elements (Originally Compiled in 1938 and Updated to 1954). – Appendix II: The 1956 Suess-Urey Abundance Table for the Individual Nuclei. – Appendix III: The 1965 Abundance Table Compiled by Virginia Trimble. – Subject Index.

Springer-Verlag
Berlin
Heidelberg
New York
Tokyo

GPSR Compliance

The European Union's (EU) General Product Safety Regulation (GPSR) is a set of rules that requires consumer products to be safe and our obligations to ensure this.

If you have any concerns about our products, you can contact us on ProductSafety@springernature.com

In case Publisher is established outside the EU, the EU authorized representative is:

Springer Nature Customer Service Center GmbH
Europaplatz 3
69115 Heidelberg, Germany

Batch number: 10704763

Printed by Printforce, the Netherlands